CHEMICAL COMPOSITION OF
EVERYDAY PRODUCTS

John Toedt, Darrell Koza,
and Kathleen Van Cleef-Toedt

GREENWOOD PRESS
Westport, Connecticut · London

Library of Congress Cataloging-in-Publication Data

Toedt, John.
   Chemical composition of everyday products / John Toedt, Darrell Koza, and Kathleen
   Van Cleef-Toedt.
      p. cm.
   Includes bibliographical references and index.
   ISBN 0–313–32579–0 (alk. paper)
      1. Chemistry, Analytic. I. Koza, Darrell.   II. Cleef-Toedt, Kathleen Van.   III. Title.
   QD75.22.T64   2005
   543—dc22   2004027860

British Library Cataloguing in Publication Data is available.

Library of Congress Catalog Card Number: 2004027860
ISBN: 0–313–32579–0

First published in 2005

Greenwood Press, 88 Post Road West, Westport, CT 06881
An imprint of Greenwood Publishing Group, Inc.
www.greenwood.com

Printed in the United States of America

The paper used in this book complies with the
Permanent Paper Standard issued by the National
Information Standards Organization (Z39.48–1984).

10 9 8 7 6 5 4 3 2 1

While the "laboratory exercises" listed in this book's appendix have been exhaustively
tested for safety and all attempts have been made to use the least hazardous chemicals
and procedures possible, the authors and publisher cannot be held liable for any injury
or damage which may occur during the performance of these "exercises."

Contents

Preface

Welcome to the enormous world of chemical products for the average consumer! We all reside in an environment richly filled with numerous consumer choices, in which chemistry plays a pivotal role. Chemistry, often described as the systematic study of all substances, is an extremely important part of human society. The study of chemistry encompasses the composition and properties of all matter and any changes exhibited when there is a reaction with other types of substances. Matter, of course, is any substance that occupies space within the universe and has a property called mass. Any substance that can be characterized and quantified by weight has mass. A change that may result from the interaction of two or more substances is defined as a chemical reaction, and the substances individually are termed chemicals. Thus, chemistry is part of the study of the functioning and properties of the physical world. In other words, chemistry is a specialization in the general field of science. Life is frequently described as a series of choices within the overall chemistry of the world. Therefore, a description of the chemistry of everyday products is simply a guide to specific types of chemical items available in modern societies that have come to symbolize convenience, prosperity, and an increase in the overall quality of our lives. The materials used by the chemical industry are chosen for many reasons, including cost, product performance, consumer use comfort, and so on. The goal of this book, although certainly not exhaustive, is to describe in general language the specific chemical composition of a range of consumer products while also highlighting such topics as the historical use of the product, physical functioning, and any associated environmental and/or human health-related issues.

Much of the fun related to the field of chemistry is derived from the creation of novel chemical formulations and/or products. As a synthesis-based science, chemistry as applied to industry is involved with virtually all aspects of our lives. A primary role of the chemical industry is to manufacture materials deemed necessary for our society, thereby allowing humans to maintain a particular type of lifestyle, to which many have become very accustomed. Although our world is a world of chemicals, it also consists of choices related to chemical products. A general knowledge of chemical concepts allows individuals to become aware of the benefits of some chemical products and the drawbacks and potential dangers of other products. Modern chemists can routinely transform "raw" materials, such as salt into lye and laundry bleach, and crude oil into cleaning detergents, drugs, lawn pesticide treatments, and a vast array of other products. The modern consumer is therefore faced nearly every day with choosing among literally hundreds of thousands of common everyday products. A trade or brand name can be given to a chemical or a formulation of chemicals and registered for its exclusive use. However, the trade name itself relays no information, and the composition of the product to which it refers can vary tremendously. The list of chemical ingredients on products, while provided by manufacturers in many cases as a result of government regulation, can appear to be written in an unfamiliar and/or "foreign" language. The consumer is thus potentially left without the knowledge-based foundational "tools" to assist him or her in understanding the actual composition, functioning, and ramifications of product choices. Although clever multimedia advertising techniques are used by many manufacturing industries to market these products, the knowledgeable and unbiased consumer is the best protection against the uncertainty and the unknown within the vast world of chemical product choices.

As this seemingly overwhelming burden of choice is entirely ours, it can be helpful to have a guide and a general description of the actual chemical composition of the products we have come to incorporate almost seamlessly into our existence. The goal of this book is to lay a very basic foundation of understanding for the general consumer interested in gaining insight into the chemical world of particular everyday products. The authors' goal was to present each product as an individual chemical entity, with an associated description of the general history of its formulation and/or use (when applicable), chemical composition, unique chemical properties (if any), and function of the major active ingredients for each product. Chemical formulae provided for various chemicals described are provided to serve as a symbolic representation of the composition and structure of molecules and compounds. Products were chosen by the authors based on a general assessment of items normally available in popular consumer retail markets throughout the United States. Such products are described in general chapters with the headings "Soaps and Laundry Products," "Cosmetics and Bathroom Products," "Health and Medical Products," "Baby Products," "Cleaning Products," "Lighting," "Common Household and Lawn Products," "Automotive and General Repair Products," and "Common Materials and Office Supply Products." Not all products within these basic categories normally available as consumer items are addressed within this text. However, the authors have tried to present a fair selection of "everyday" consumer products currently marketed, which are in most cases used literally every day, used somewhat frequently, or will probably be used by many consumers at some point in their lives. In addition, a selection of general laboratory exercises is provided for the reader at the end of the text to allow for a better appreciation of some basic concepts and techniques practiced by many individuals training to become active members of the chemical profession.

We hope that this text provides the inspiration for learning about chemistry, in that it will allow for the proliferation of what have been come to be called "educated consumers." Such consumers are the foundation of the freedom concept of consumer choice and often provide the chemical industry with ideas for the manufacture of better products as well as ideas for better product uses within our ever-changing society and often fragile natural environment. Chemical knowledge thus enables consumers to thoroughly understand and enhance the choices made during their lifetimes.

1

Soaps and Laundry Products

For a substance to play a role in cleaning, it must remove items such as dirt and grease from various materials (e.g., clothes, carpets, surfaces, animal skin tissue, etc.). Dirt and grime usually adhere to skin, clothing, and other types of surfaces because they combine with greases and oils (e.g., body oils, cooking fats, lubricating greases, etc.) that act similarly to glues. Because oil is not hydrophilic and thus not soluble in water, attempting to wash dirt and grime from surfaces using water alone results in little success. Soaps and detergents usually found in a typical modern consumer home can be grouped into four categories: personal cleaning/hygiene, laundry, dishwashing, and household cleaning. In all types of cleaning, the soil is transferred from the dirty object into a separate phase, either liquid (e.g., water) or solid, and then removed from the object. Essential to personal and public health, soaps and detergents contribute to good personal hygiene, reduce the presence of disease-causing microorganisms, extend the value and usefulness of clothing and household surfaces and furnishings, and maintain environments of well-being, all through their ability to loosen and remove soil from various synthetic and natural surfaces.

To understand the physical process involved with effective cleaning, it is helpful to understand basic soap and detergent chemistry. One of the most common liquids used for cleaning is water. Water (H_2O) has a property called "surface tension." Within/throughout the bulk of the liquid water, the total force of attraction exerted by any one molecule of water on all its surrounding neighbors is dispersed spherically, in all directions. However, at the surface, where there are only occasional molecules of gases in the atmosphere and the very few water molecules that have escaped via evaporation, the water molecules are only surrounded by other water molecules below and to the side. A tension is created as the water molecules at the surface are pulled into the main body/bulk of the liquid water. It is this strong attraction of the surface water molecules for each other and for the water molecules immediately below them that results in the cohesive property of the surface known as surface tension. Surface tension causes water to "bead up" on surfaces (e.g., human skin, glass, cloth fabric, furniture), which slows the overall wetting of the surface and thereby inhibits the cleaning process. For the cleaning process to be successful, the surface tension must be reduced so that water can evenly overspread, wet, and saturate surfaces. Organic (carbon-based) chemicals that decrease surface tension are termed "surface active agents" or, simply, "surfactants." When added to water, the hydrocarbon chains of surfactants form the uppermost layer in the liquid. While these hydrocarbon chains pull together and create surface tension, they attract one another with what are called van der Waals forces, not hydrogen bonds, and thus create a surface tension only about 30 percent that of water molecules. Overall, the presence of surfactants within a body of water significantly reduces the surface tension normally exhibited by the water.

Surfactants are also involved with other aspects of successful cleaning, including loosening, emulsifying (dispersing one liquid into another immiscible liquid, which prevents oily soils from resettling on the surface of water), and holding soil in suspension until it can be removed and/or washed away. The arrangement of soap micelles within water is called a stable emulsion. Unlike a regular mixture of oil and water, this arrangement does not separate when it sits unattended. Thus, surfactants may act as emulsifiers, substances that help to form and stabilize emulsions. This means that, although oil, which attracts greasy dirt, does not actually mix with the water, soap can suspend the oil and greasy dirt in such a way that it can be removed. Therefore, the cleaning power of many surfactants results from the enhanced ability of the water to wet the normally hydrophobic (non-water-soluble) surface, penetrate into fibers more freely, and lift off the dirt. For this reason, surfactants are often referred to as "wetting agents" because they help water actually wet surfaces. Many surfactants provide an alkaline environment within the cleaning process, which can be essential in removing acid-based soils and grime.

Surfactants are classified by their ionic (electrically charged particles) properties in water. Surfactant molecules can be described as resembling a tadpole (immature frog) because they contain a fairly long fatty acid tail (hydrophobic or water insoluble) and a small, often electrically charged head (hydrophilic or water soluble). The long hydrocarbon (CH_2 groups) tails of surfactants are soluble in hydrophobic substances such as oil, and the hydrophilic heads of surfactants containing carboxyl or sulfonate groups are soluble in water. Water is polar: the H_2O molecules have an attraction for other polar substances, such as common table salt (NaCl). When salt is added to water, the salt molecule ions are attracted to, and become surrounded by, the water molecules. This occurrence is known as solubility. Oil, however, is a nonpolar substance and therefore will only dissolve in other nonpolar substances. Because nonpolar substances cannot form hydrogen bonds with water molecules, they do not bind well with water and are essentially insoluble in water. Thus, there is an actual chemical foundation behind the phenomenon that oil and water do not mix! Surfactants promote an environment in which oil can seemingly dissolve in water by bridging the oil-water interface. There are four possible combinations of surfactants: (1) anionic, in which a negative charge is concentrated in the hydrophilic head region; (2) nonionic, which does not have a specific charge; (3) cationic, in which the hydrophilic head region carries a positive charge; and (4) amphoteric, which carries both a positive charge and a negative charge in the same molecule. Amphoteric surfactants act as either an anion (negative charge) or cation (positive charge), depending on the pH of the solution in which they are used. Many cleaning products contain two or more types of surfactants to optimize their intended cleaning use.

Detergents are excellent examples of surfactants. The word "detergent" comes from the Latin *detergere*, meaning to clean or to wipe off. Thus, a detergent is any chemical substance that cleans, particularly if the detergent removes nonpolar substance such as greasy oils, fats, and waxes from skin, food, and plants. Typical modern detergents include soaps and synthetic detergents. As the water surface becomes saturated with detergent surfactant molecules and additional detergent is added, the excess detergent molecules will become crowded out of the surface. When the concentration of detergent molecules in the water reaches a certain value, called the "critical micellar concentration," the excess detergent molecules initiate the process of shielding their hydrophobic tails from the water molecules by clustering into micelles within the main body/ bulk of the liquid water. In general, the term "micelle" refers to very small (submicroscopic) spheres or globules of a particular substance distributed throughout another substance, usually a liquid. These communal micelle molecules contain roughly forty to one hundred molecules. Within the bulk of the liquid water, the hydrophilic heads of the detergent molecules form the surface of the spheres, while the hydrophobic tails point inward, forming a group of hydrophobic hydrocarbon chains shielded from the surrounding water molecules. Thus, the detergent molecules are well dispersed in the water, present as a colloid, but are not actually dissolved. The inside of the micelle is nonpolar and therefore tends to collect oily soil molecules. Greasy oils and dirt within the water dissolve in the hydrophobic tail portion in the center of the micelle and are broken down into tiny droplets and dispersed within the aqueous solution. The micelles float freely within the body of water and collect and hold onto any oil molecules they encounter. The oily dirt is then washed away, possibly down the drain, as a result of the interaction/attraction of the hydrophilic head portion with the water molecules on the surface of the micelle spheres. Polar soil molecules, including salts from perspiration and ground dirt, simply dissolve in the water, where they become ions that are carried away in the shells of water molecules.

Soap, other detergents, and most dirt are biodegradable. For example, when a detergent/dirt/water suspension is released into the environment via a drainage system, it can be converted into either biomass or carbon dioxide and the associated remaining water returned to the natural environment.

PERSONAL CLEANSING PRODUCTS: BAR SOAP

Soap is the oldest surfactant; it is thought to have been in use for more than 4,500 years. The origins of personal cleanliness quite possibly date back to prehistoric times. Early people cooking their meats over fires may have noticed that after a rainstorm, the strange foam around the remains of the fire and its ashes caused their cooking pots and hands to become cleaner than was ordinarily expected. In early societies that developed near waterways, a soaplike substance is thought to have been extracted from plants such as soapwort, soap root, soap bark, yucca, horsetail, fuchsia leaves, bouncing bet, and agave, all of which tend to flourish on riverbanks or near lakes. It is recorded that Babylonians were making soap around 2800 BC. Evidence of such soap making was made apparent after a soaplike material was found in clay cylinders during an excavation of ancient Babylon. Inscriptions on the cylinders indicated that fats were boiled with ashes, which is a known method of soap making. Evidence also indicates that soap making was known to the Phoenicians around 600 BC. While the evidential purpose of this early "soap" was unclear, it is thought that these early references to soap and soap making indicate the use of soap in the cleaning of textile fibers (e.g., cotton and wool) in preparation for weaving cloth, and later as hair-styling products or as a medicament on wounds. Evidence also indicates that Egyptians bathed on a regular basis, and the Ebers Papyrus (a medical document dated approximately 1500 BC) describes a soaplike material synthesized from combining animal and vegetable oils with alkaline salts, which was used for both washing and the treatment of skin diseases. Also at this time, Moses provided the Israelites with detailed laws concerning personal cleanliness and health, and biblical accounts indicate that the Israelites possibly knew that hair gel was produced by combining oil and ashes. While the early Greeks bathed for aesthetic reasons, they chose to clean their bodies with blocks of clay, sand, pumice, and ashes rather than with soap. A metal implement known as a "strigil" was used to scrape off oils and ashes used to anoint bodies, and body dirt apparently was removed with this scraping process.

In ancient Rome, oils, unguents, plant essences, and cosmetics were apparently used in heavy quantities, but there is no reference to soaps and their use as cleaning agents. While the Romans were known for their practice and use of public baths, personal cleaning involved rubbing their bodies with olive oil and sand and using a strigil to scrape the oil, sand, dirt, grease, and dead skin cells off their bodies. However, the name "soap" is thought to have originated, according to an ancient Roman legend, from Mount Sapo, where animals were sacrificed. Rain poured down this mountain, through a mixture of melted animal fats, or tallow, and wood ashes into the clay soil along the Tiber River below. Women washing clothes at the river apparently noticed that clothing exposed to the soapy mixture of saponified acids (fats) and alkali (caustic ashes) within the river water become cleaner very quickly with little effort. Saponification, the chemical term for the "soap-making" reaction, bears the name of this mountain in Rome. The first reliable evidence of soap making is found in the historical accounts of ancient Rome. The Roman historian, Pliny the Elder, described the synthesis of soap from goat tallow and caustic wood ashes and also indicated that common salt was added to harden soap. The Romans knew, long before the actual chemical process was completely understood, that heating goat fat with extracts of wood ashes, which contain alkaline (basic) products (e.g., potassium hydroxide [KOH] and potassium carbonate [K_2CO_3]), produces soap. The first reaction formed potassium hydroxide, which causes the breakdown of the fat triglycerides into the component parts, glycerine and fatty acids. In the process, the fatty acid is neutralized by the strong alkali and ends up in the salt form. The Romans also used lye (sodium hydroxide [NaOH]), a stronger base than ash extracts, and more effective in changing fats into actual soap. The word "lye" is apparently related to soap and the process of soap making through an extensive path of linguistics, including words from Latin, Greek, Old English, Old Irish, and other languages, meaning lather, wash, bathe, and even ashes. In AD 79, the city of Pompeii, Italy, was destroyed after the eruption of the volcano named Mount Vesuvius. Interestingly, excavation of Pompeii revealed an entire soap-making factory, complete with finished bars of soap preserved in the hardened lava. Soap used as a personal cleansing technique had become popular during the later centuries of the Roman Empire.

By the second century AD, the Greek physician named Galen was recommending soap for both medicinal and bathing/cleansing purposes. Although the first Roman baths had been built in approximately 312 BC, supplied with water from their extensive aqueducts, the fall of the Roman Empire in AD 467 resulted in the decline in bathing habits in Western Europe. There was little soap making performed or use of soap for cleaning in the European Dark Ages. While there were apparently public bathhouses, called stews, where patrons were provided with bars of soap for personal cleansing during the European Middle Ages, it was later during medieval times when bathing fell out of custom. Many historians suspect that the lack of personal cleanliness and related unsanitary conditions substantially contributed to the outbreak of the great disease plagues of the Middle Ages, especially the occurrence of the Black Death of the fourteenth century. Remaining stews were closed, as authorities suspected they promoted the spread of disease. It is well known that during the Renaissance, people preferred to cover their bodies with heavy scents rather than try to maintain body cleanliness. However, daily bathing was a common custom in Japan, and weekly hot spring bathing was popular in Iceland during the Middle Ages.

The Celtic peoples are also thought to have discovered soap making. Many historians believe that, possibly because of increased contact with the Celts by the Romans, soap was used by the Celtic people for personal washing. It is also possible that the Celts and the Romans independently discovered saponification. In the Byzantine Empire, the cultural remains of the Roman Empire in the eastern Mediterranean region, in the expanding Arab nations, and in the regions conquered by the Vikings, soap was manufactured and used. The ancient Germans and Gauls are also credited with discovering a substance called soap, made of tallow and ashes, that they used to dye their hair a reddish color. Soap making was an established craft in Europe by the end of the seventh century. Soap making was revived in Italy and Spain, beginning in approximately the eighth century. By the thirteenth century, the city of Marseilles, France, emerged as a prominent soap-making center for the European markets. During the fourteenth century, England also formally initiated saponification techniques. At this time, saponification usually involved combining vegetable and animal oils with ashes of plants, along with pleasant fragrances. While soaps produced in southern Europe, Italy, Spain, and southern ports of France were synthesized from high-quality olive oils, soaps produced in England and northern France were produced from lower-quality animal fats (e.g., tallow, the fat from cattle), and soaps produced in northern Europe often resorted to fish oils. Southern European countries enjoyed a lively trade of exporting fine soaps, as these regions all boasted a rich supply of olive oil as well as barilla, a fleshy plant whose ashes were used to make the lye formula required for saponification at that time. Soap-making guilds and trade associations recognized the need to regulate the lucrative soap-making process, and soaps were taxed as luxury items in England and France throughout the seventeenth, eighteenth, and nineteenth centuries. The habit of bathing came back into fashion and the consumption of soap increased tremendously in the nineteenth century. Abolishment of some European soap taxation eased the transition of soap from a luxury item to a common personal hygiene household commodity.

While the first European settlers to New England carried soap with them on the ships chartered by the Massachusetts Bay Company and later manufactured soap themselves from the by-products (e.g., boiling of wood ash lye, cooking grease and animal fats together) of their home-steading activities, commercial soap making in the American colonies began in 1608 with the arrival of several Polish and German soap makers on the second ship from England to reach Jamestown, Virginia. Profes-sional soap makers who traded and sold soap, often called soapboilers, would collect large amounts of waste fats from households in exchange for small amounts of presynthesized soap. Since tallow was also the main ingredient for candles, many soapboilers produced both soap and can-dles. A major step toward large-scale commercial soap making occurred in 1791, when a French chemist, Nicholas Leblanc, patented a process for making soda ash (lye), or sodium carbonate, from a brine solution of common salt. Soda ash is the alkali obtained from ashes that is combined with various fats to form soap. The Leblanc process was an easy, inexpen-sive technique that yielded large quantities of good-quality soda ash at the industrial level. During the 1800s, other French and Belgian chem-ists contributed to the advancement of soap technology by discovering the chemical nature and relationship of fats, glycerine and fatty acids, and by inventing the ammonia process, which uses common table salt (so-dium chloride [NaCl]) to make soda ash inexpensively and in increased quantities for quality high-volume soap production. Thus, during Victo-rian times, soap came of age when saponification turned from part craft and folklore item to a fully developed industry. Soap making was one of America's fastest-growing industries by 1850, with the rise of companies such as Colgate & Company, Palmolive, Lever Brothers (an English company), and Procter & Gamble. The broad availability of soap allowed for the product to change from a luxury item, frequently synthesized as a household art, to an everyday necessity item with widespread use.

Soaps actually make up a very narrow class of detergents. The term "soap" is restricted to the sodium (or, infrequently, potassium) salts of long-chain carboxylic acids. As indicated above, natural soaps, the sodium or potassium salts of fatty acids, were originally synthesized via saponification, a process of boiling lard or other animal fats (and later vegetable fats) together with lye or potash (potassium hydroxide). The fats and oils thereby undergo hydrolysis, yielding crude soap and a by-product called glycerol. An example of such a crude soap is sodium octadecanoate ($C_{17}H_{35}COONa$). Soaps are anionic (negatively charged) surfactants produced from the hydrolysis of fats in a chemical reaction called saponification. In other words, soaps are water-soluble sodium or potassium salts of fatty acids synthesized from fats and oils (or their fatty acids) by chemically treating them with a strong alkaline substance (base). Each soap molecule has a long hydrocarbon chain (CH_2 groups that terminate in a CH_3 group), often referred to as a "tail," that is linked by a covalent bond to a carboxyl group (CO_2H), often called a "head." The anion of the carboxyl group is balanced by the positive charge of a sodium or potassium cation. In water, the sodium or potassium ions float freely, leaving the head region with a negative charge. It has an oxygen end, which is polar, and a fatty acid end, which is nonpolar. Thus, part of the molecule contains a polar hydrophilic (water-loving) structure; the ionic head region is attracted toward water molecules but shuns hydrocarbons and other oil-based substances. The long hydrocarbon tail of the molecule is a nonpolar hydrophobic (water-fearing) structure; it shuns water but mixes easily with greasy, oil-based substances that repel the hydrophilic head portion. The hydrocarbon chains of various soap molecules within a body of water are attracted to each other by dispersion forces and cluster together to form micelles. Within these micelles, the carboxylate head groups form a negatively charged spherical surface, with the hydrocarbon chains inside the sphere attracting oily dirt. Because the outer portion of the micelles are negatively charged, the soap micelles repel each other and remain dispersed in water. Grease and dirt are attracted to the nonpolar hydrocarbon portion of the micelles, are subsequently "caught" inside the micelles, and then can be rinsed away.

Soaps are detergents that both lower the surface tension of water, which allows water to penetrate a dirty substance and then suspend the grease droplets within the micelles, preventing them from redepositing on the substance being cleaned, and allow the greasy dirt to be washed away. Since the anionic carboxylate groups of the soap molecules provide a cover of negative electrical charge on the surface of each micelle, soaps are considered anionic detergents.

The fats and oils used in soap making originate from animal or plant sources. Each fat or oil is composed of a distinctive mixture of several triglycerides. Triglycerides, the principal organic compounds of animal and vegetable oils, are composed of three fatty acid molecules attached to one molecule of glycerine. Fatty acids are the components of fats and oils that are used to synthesize soap. They are weak acids composed of two parts: a carboxylic acid group and an attached hydrocarbon chain. Because of the combination of both hydrophilic and hydrophobic properties in a soap molecule, it is necessary to obtain a balance between the two properties to achieve the greatest benefit. For example, the best benefit seems to be derived from the reference of the C-14 fatty acid molecule within the hydrocarbon chain. As the hydrocarbon chain is lengthened, the solubility in water decreases. Above C-18, the solubility becomes too low to be of any practical use. When the hydrocarbon chain is shortened, the ability to suspend oil droplets is significantly decreased. Sodium salts of fatty acids containing ten to eighteen carbons make the best soaps. There are many types of triglycerides, with each type possessing a particular combination of fatty acids. Coconut oil is considered quite suitable because of the unusually high content of lauric, myristic, and palmitic acids. Tallow, fat from beef and lamb, is predominantly composed of palmitic, stearic, and oleic acids. Castile soap is made primarily from olive oil, which can be less irritating to the skin and prized by some people for that reason. The alkalis used in soap making were originally obtained from the ashes of plants but are now available as commercially synthesized products. An alkali is a soluble salt of an alkali metal such as potassium or sodium, but the term can also refer to a substance that is a base, which reacts with and neutralizes an acid to form a salt. The common alkalis used in soap making are sodium hydroxide, called caustic soda, and potassium hydroxide, also called caustic potash.

Saponification, the process of soap making, occurs when fatty acid-containing fats and oils are mixed with a strong alkali. This method involves heating fats and oils and subsequently reacting them with a liquid alkali to produce soap and water plus glycerine. In the industrial manufacture of soap, tallow (fat from beef or lamb) or vegetable fat (e.g., coconut oil, olive oil, etc.) is heated in the presence of sodium hydroxide. Once the saponification reaction is completed, a salt (e.g., sodium chloride) is added to precipitate the soap from solution. The water layer is then removed from the top of the mixture and the glycerol by-product is recovered, often using vacuum distillation. Impurities within the crude soap (e.g., sodium chloride, sodium hydroxide, and glycerol) are subsequently removed by repeating the boiling and salt-precipitation processes. Another major soap-making process involves the neutralization of fatty acids with an alkali. Fats and oils are hydrolyzed with a high-pressure steam to yield crude fatty acids and glycerine. Hydrolysis (the decomposition of a substance, or its conversion to other substances, through the action of water) of the triglyceride generates both glycerol, which remains dissolved in the water, and the salts of the various acids that make up the triglyceride. The fatty acids are then purified by distillation and neutralized with an alkali to produce soap and water. After the splitting of the fats and oils, the sodium or potassium portion of the alkali joins with the fatty acid molecules. The salts of the fatty acids actually congeal at the surface as the mixture cools, and these acids constitute the soap. Historically, both the glycerol and the soap contained unreacted alkali, which destroyed skin tissues with each use, thus substantiating the deleterious effects caused by improperly homemade "lye soap." When the alkali added is sodium hydroxide, sodium-based soap is formed. Such sodium soaps are "hard" soaps. Glycerine is also considered a valuable by-product and can be recovered by chemical treatment, followed by evaporation and refining. Refined glycerine is a primary ingredient of many soaps available over the counter and as an important industrial material used in such items as foods, cosmetics, and drugs.

Bar soaps are formulated for cleaning the hands, face, and body. Traditional bar soaps are made from fats and oils (which are largely esters) or their fatty acids reacted with inorganic water-based bases. They have a general formula, $RCOO^-Na^+$, where R is a long hydrocarbon chain, $CH_3(CH_2)_{10-16}$. The saponification reaction is summarized as follows: an ester (fat) and a base (caustic soda, sodium hydroxide) react together to yield the salt of a fatty acid (soap) and an alcohol (e.g., glycerol). This is simply the reverse of an esterification reaction. The main sources of fats include beef and lamb (mutton) tallow and white palm, coconut, and palm kernel oils. These raw materials are pretreated to remove impurities and to achieve the color, odor, and performance features desired (e.g., sudsing ability) in the finished soap bar product. The hardness of soap depends on the fats from which it was made. Saturated fats (fats that remain fairly solid at room temperature, such as those derived primarily from animal fats) produce the hard bar soaps. Sodium palmitate, the sodium salt of palmitic acid, is a typical soap. Beef tallow yields principally sodium stearate $[CH_3(CH_2)_{16}COO^-Na^+]$, one of the more common soaps. Palm oil yields sodium palmitate $[CH_3(CH_2)_{14}COO^-Na^+]$, a component in many expensive soaps. Standard soaps comprise approximately 80 percent tallow and 20 percent coconut oil, with added chemical sudsing agents. The total amount of water permissible is usually 17 percent. The characteristic of good lathering/sudsing ability is attributable to the fact that as surfactants, soap molecules tend to align along the surface of the water, with the hydrocarbon chains directed toward the surface and the saltlike ends directed into the solution exposed to the water. The water surface is thus weakened and promotes foaming. The density of soap can be decreased by the incorporation of air to allow for floating. Additives within modern soaps include creams (emollients), perfumes, preservatives, antioxidants, deodorants, abrasives (e.g., pumice, silica), and colorings (e.g., titanium dioxide in the case of white soaps). Soaps termed "combo bars" usually are formed from a combination of actual soap and synthetic surfactants. Specialty bars include transparent/translucent soaps, luxury soaps, and medicated soaps.

Although soaps are excellent cleansers, their effectiveness is limited when used in hard water. Hardness in water is attributable to the presence of various mineral salts, including calcium (Ca), iron (Fe), magnesium (Mg), and manganese (Mn). The mineral salts react with soap to form insoluble salts. This collective insoluble precipitate is often referred to as "soap film" or "soap scum." Hard water wastes soap because a good portion of the soap that would otherwise be used for cleaning is consumed in precipitate formation as it reacts with the mineral ions of hard water. In addition, as salts of weak acids, soaps are converted by mineral acids into free fatty acids, and these fatty acids are less soluble than the sodium or potassium salts. Thus, soaps are also ineffective in acidic water. The insoluble salts form bathtub rings, leave films on skin, and gray/roughen bathroom tiles with repeated washings. However, soap is an excellent cleanser in soft water, it is relatively nontoxic, it is derived from renewable resources (animal fats and vegetable oils), and is biodegradable.

SYNTHETIC DETERGENTS: SURFACTANTS

In the modern world, the laundering of clothes and other types of cleaning processes have been accomplished via the displacement of soap in favor of products called synthetic detergents. Laundry detergents are currently available as liquids or powders and are formulated to perform soil and stain removal and fabric treatment (e.g., bleaching, softening, and conditioning) under varying water chemistry conditions and temperatures. While liquid detergents are recommended for use on oily soils and the pretreatment of soils and stains, powders are apparently quite effective in lifting clay-based soils and ground-in dirt off fabrics. Laundry detergents first came into use because of problems encountered with ordinary soaps, including deterioration on storage, lack of specialized cleaning ability, and lack of complete rinsing out of fabrics after washing. In addition, soaps were unsatisfactory for laundering in hard water. The characteristic of water "hardness" can be attributed to the presence of metal ions (excluding group 1A metals), primarily calcium (Ca^{2+}) and magnesium (Mg^{2+}) ions, which form insoluble precipitates when combined with the fatty acid anions of ordinary soap. An example of such a precipitate is the formation of the very undesirable "bathtub ring" within the bathtub or washbasin. If soap were used as a primary laundering agent, such precipitate residues would also gradually build up as deposits on clothing, causing bad odor and the deterioration of fabric materials. Thus, the first synthetic detergents were developed in an effort to overcome the reaction of soaps with hard water. While the chemical structure of synthetic detergents differs slightly from that of soaps, they are both efficient surfactants.

The chemical process of soap synthesis remained steadily similar throughout the ages, until approximately 1916, when the first synthetic detergent was developed in Germany in response to the shortage of available fats for soap making during World War I. These detergents were basically short-chain alkyl naphthalene sulfonates, made by coupling propyl or butyl alcohols with naphthalene and subsequent sulfonation (addition of sulfur atoms). In the late 1920s and early 1930s, long-chain alcohols were sulfonated and sold as the neutralized sodium salts. Household detergent production in the United States began in the early to mid 1930s but did not increase in manufacture until after World War II, when the raw materials for soap manufacture were both scarce and costly. These detergents were long-chain alkyl aryl sulfonates with benzene as the aromatic center and the alkyl portion synthesized from kerosene. During World War II, fat and oil supplies were interrupted, and the need for a cleaning agent for use in mineral-rich cold seawater was realized within the U.S. military. Further research concerning detergent chemistry was stimulated, and some of the first commercially available synthetic detergents were marketed/used for hand dishwashing and the laundering of fine fabric. Within a few years, cheap synthetic detergents, mostly synthesized from petroleum products, were widely available. During this time, alkyl aryl sulfonates had nearly completely overshadowed the sales of alcohol sulfate-based laundry detergents. Molecules of synthetic detergents were similar enough to soap to have the same excellent cleansing action but sufficiently different to resist the effects of acidic and hard water.

The ingredients in any commercially available synthetic detergent are contained within eight different groups known as the surfactant system: the builders (both inorganic and organic), fluorescent dyes (optical whiteners), enzymes, corrosion inhibitors, bleaches, a filler, fragrances, and coloring agents. While builders, bleaches, optical whiteners, and enzymes will be treated as laundry aids as separate everyday products, the functions of fillers, corrosion inhibitors, fragrances, and coloring agents will be discussed here. One of the major functions of fillers (also called processing agents), primarily sodium sulfate (Na_2SO_4), was/is to add bulk to the detergent to provide the consumer with a fair sense of volume verses monetary product worth. Fillers also allow granular detergents to pour from the packaging box more freely. The amount of filler within a given detergent product may range from 35 percent to 0 percent. Corrosion inhibitors prevent the potential deleterious effects of detergent ions that would otherwise quickly rust the steel inside a washing machine. Rusting is an electrochemical process whereby the iron in steel is attacked by negatively charged hydroxyl ions, or detergent ions. These compounds are usually sodium silicates, water-soluble glasses. Fragrances add pleasant odors to the detergent and thus to just-washed fabrics. Coloring agents, primarily blue-toned coloring agents, coat fabrics and provide a blue tint to washed fabrics. Bluing agents contain a blue dye or pigment taken up by fabrics in the wash or rinse cycles; thus, bluing absorbs the yellow part of the light spectrum, counteracting the natural yellowing of many fabrics. This process adds a small amount of brilliance to white fabrics and gives them an appearance of extra cleanliness.

The surfactant system includes major active ingredients such as detergents, which form micelles and clean grease off clothing and other items through the same mechanism as soaps. Warm or hot water assists in dissolving grease and oil within soiled clothing, and modern washing machine agitation (or hand rubbing) helps lift soil out of fabrics. Modern detergent surfactants are synthesized from a variety of petrochemicals (derived from petroleum products) and/or oleochemicals (derived from fats and oils). Similar to the fatty acids used in soap making, petroleum, fats, and oils contain hydrocarbon chains that are hydrophobic but attracted to greasy oils and dirt. The hydrocarbon chains are the foundation of the hydrophobic tail end of the surfactant molecule. Other chemicals, including sulfur trioxide, sulfuric acid, and ethylene oxide, are used to produce the hydrophilic head end of the surfactant molecule. As in soap making, an alkali is used to make detergent surfactants, with sodium and potassium hydroxide being the most common alkalis used.

The solution for the soap precipitation problem in hard water was first realized with the development of alkylbenzene sulfonate (ABS) detergents. ABS detergents were synthesized from propylene (CH_2=$CHCH_3$), which is available from petroleum, benzene, and sulfuric acid (H_2SO_4). Alkylbenzene is a product of the petroleum industry and is made by the condensation of an α-olefin with benzene. By treating this insoluble material through a process called sulfonation (adding an excess of sulfuric acid), it is converted to the corresponding sulfonic acid. The resulting sulfonic acid (alkylbenzenesulfonic acid [RSO_3H]) is then neutralized with a base (e.g., sodium carbonate [$NaCO_3$] or sodium hydroxide [$NaOH$]), to yield the final branched-chain product (e.g., ABS). The similarities of detergent ($RSOOO^-Na^+$) and soap ($RCOO^-Na^+$) are obvious. Thus, soap molecules have a carbon (C) atom bonded to the oxygen to which the sodium is bonded, and detergents have a sulfur (S) atom in that position. Similar to soap, this ABS synthetic detergent is both hydrophilic and hydrophobic. The detergents produced in this manner are significantly more soluble than soap. Thus, it can stabilize water-oil emulsions, but unlike soaps (fatty acid salts), the sodium can be replaced by calcium or magnesium and the detergent remains in solution, even in hard water (i.e., soap scum is not formed). However, ABS detergents are significantly more stable than soaps and can persist in wastewater systems long after use and discharge.

The increased stability of ABS detergents results from the greater chemical stability of the sulfonate grouping and the raw material petroleum-based branching characteristic of the long hydrocarbon chain molecules, in sharp contrast to the straight-chain hydrocarbons derived from animal fats. In the natural environment, bacteria did not readily degrade branched ABS detergents; thus, foaming and sudsing began to accumulate within sewage treatment plants, natural waterways (e.g., rivers), and even some reservoirs used as sources of drinking water. By the early 1960s, groundwater supplies were perceived as threatened, and public outcry, along with governmental legislation, persuaded the United States to solve the problem by replacing ABS detergents with biodegradable linear alkylbenzene sulfonate (LAS) detergents. This detergent consists of a long hydrocarbon chain attached to an aromatic or benzene ring attached to a negatively charged sulfonate group. The sulfonate group involves one sulfur atom and three oxygen atoms. The negatively charged head structure allows for the entire surfactant molecule to be easily carried away by water molecules. LAS detergents possess linear unbranched chains of carbon atoms within the hydrocarbon tail of the molecule, often referred to as a linear alkyl group. Microorganisms can thus readily break down LAS molecules by producing enzymes that degrade the molecule, two carbons at one time. Branching of the ABS detergents had inhibited this enzyme degradation reaction. Consequently, the ABS-containing detergents, a major detergent formulation throughout the 1950s, were replaced around 1965 by detergents containing LAS surfactants, which are readily biodegradable (the process of decomposition of an organic [carbon-based] material by naturally occurring microorganisms) and non-polluting. In addition, many liquid detergent formulations contain other efficient, yet more expensive, surfactants called alcohol ether sulfates (AES). AES molecules have a hydrocarbon tail portion derived from an alcohol (or alkylphenol), a polar head portion derived from ethylene oxide, and a sulfate portion.

In domestic synthetic detergents, nonionic surfactants are increasingly used, but anionic surfactants predominate. Anionic detergents, which constitute the great volume of all synthetic powder detergents, are particularly effective at cleaning fabrics that absorb water readily, such as those manufactured from natural fibers (e.g., cotton, wool, and silk). Anionic surfactants (e.g., linear alkylbenzene sulfonate, alcohol ethoxy sulfates, alkyl sulfates) react with hydrocarbons derived from petroleum, fats, or oils to produce new acids similar to fatty acids. A second reaction adds an alkali to the new acids to produce a type of anionic surfactant molecule. Nonionic detergents, many of which have a polar end group that is not ionic and large numbers of oxygen atoms covalently bonded to their hydrophilic structures, are particularly effective in cleaning synthetic fabrics (e.g., polyester). Most nonionic detergents are used to produce liquid laundry detergents and produce little foaming action. Typical nonionic detergents have a phenolic group $[C_6H_4(OH)]$, an extremely polar but nonionic group. Nonionic surfactant molecules (e.g., alcohol ethoxylates, alkylphenol ethoxylates, coconut diethanolamide [an alkylolamide]) are produced by first converting the hydrocarbon to an alcohol and then reacting the fatty alcohol with an ethylene oxide. The alkylolamide is prepared by making the fatty acids obtained from coconut oil react with an ethylene oxide derivative called monoethanolamine. These nonionic surfactants are not actually salts and tend to be rather waxy products or liquids. These nonionic surfactants can then be reacted further with sulfur-containing acids to form another type of anionic surfactant molecule. While nonionic surfactants typically do not keep dirt particles in suspension as well as anionic surfactants, some nonionic surfactants have the unusual property of being more soluble in cold water than in hot and therefore more suitable for cold water laundering needs. Nonionic surfactants are unaffected by hard water and are actually better than anionic detergents at removing particular soils—for example, they are well suited to remove skin oils from synthetic fibers.

LAUNDRY AIDS

In general, laundry aids contribute to the effectiveness of laundry detergents and provide unique functions. Laundry aids include builders, bleaches, enzymes, fabric softeners, and optical whiteners.

Builders

Despite the considerable advances made in the production of the active detergent/surfactant chemicals starting during the early 1900s, progress in the use of detergents for heavy-duty (cotton) washing was still relatively slow. By the end of World War II, detergents had already displaced soaps to a considerable extent in the field of fine laundering and dishwashing. However, small amounts of dirt were being redeposited uniformly over the whole surface of clothing items washed in either the washbasin or the washing machine, thus giving the clothing a gray appearance. The breakthrough in the development of detergents for all-purpose laundering came in 1946, with the introduction of the first "built" detergent product in the United States. Originally created by Procter & Gamble in 1943, this detergent was a combination of synthetic detergents and "builders." Soap for cotton washing had been built for many years with various alkaline materials, including carbonates, silicates, borax, and orthophosphates. A builder is any substance added to a surfactant to increase the efficiency of detergency. Builders enhance or maintain the cleaning action of the surfactant. The primary function of builders is to reduce water hardness, and this is accomplished either by sequestration or chelation (holding water hardness minerals in solution), by precipitation (forming insoluble precipitates), or by ion exchange (trading electrically charged particles). Thus, because many consumer homes do not have water softeners installed on the premises, most modern laundry detergents soften the water themselves. In general, builders can also supply and maintain alkaline washing conditions, assist in preventing the redeposition of removed soils, and emulsify oil- and grease-based soils.

Essentially water softeners, builders continue to be included in modern domestic laundry detergent formulations to assist the surfactant system in its action. The two major types of builders include inorganic builders and organic builders. Common inorganic builders, once widely used but now universally banned or legally restricted by many governments of the world, are phosphates, specifically sodium tripolyphosphate (STPP) ($Na_5P_3O_{10}$). STPP buffers the washing water to a milder pH than would be obtained otherwise. The building process is accomplished via sequestration; specifically, STPP sequesters calcium and magnesium ions (ions that contribute to the formation of hard water) to form soluble complexes. This action softens the water and provides a mild alkaline environment, both of which result in more favorable water chemistry for optimum detergent action. STPP also allowed clay-type dirt to remain in suspension (a process called deflocculation). These phosphates vastly improved detergency performance, making synthetic detergents more suitable for cleaning heavily soiled laundry (i.e., "heavy-duty" detergency). However, the large amount of phosphates released into the environment via STPP, specifically added to lakes and streams upon discharge, was found to be the leading cause of objectionable algal blooms. Phosphates are excellent nutrients for algae and other small aquatic plants inhabiting such places as lakes and streams. When a large amount of phosphates became available, these molecules acted as fertilizers and algal growth was uninhibited, leading to a process called eutrophication (meaning nutrition by chemical means). In short, the algal blooms covered the surface of the water bodies and synthesized large amounts of oxygen via photosynthesis. However, as algal growth continued, the oxygen produced at the surface escaped into the atmosphere and prevented atmospheric oxygen from being obtained by other marine life (e.g., fish). Since algae carry out respiration and cause a net drain on the oxygen supply, the demand, called the biochemical oxygen demand, may exceed the supply and cause the destruction of much aquatic life. The problem may be accentuated as algal blooms block out sunlight required by other photosynthesizing aquatic plants. In addition, the death of the overwhelmingly high populations of algae over time, with their associated settling to the bottom of the water bodies, allowed for extensive tissue decomposition by bacteria, the emission of obnoxious sulfurous odors, and thus further depletion of oxygen supplies. It was this death of fish and other aquatic animals, often occurring on a large scale in lakes and rivers covered by algae, that prompted environmental action through government legislation at the height of the phosphate controversy in the early 1970s. Ironically, the problem was actually found to be related to the total amount of detergent used by consumers, and the level of aquatic pollution, not the actual phosphate. Unpolluted lakes can readily absorb excess phosphate, as zooplankton will eat the algae that flourish and are themselves eaten by fish. However, the addition of phosphates to polluted waters leads to unchecked/unbalanced algal growth. In mid 1970, the major detergent producers announced plans to replace a large percentage of the phosphate as STPP (20 to 65 percent in detergent formulations) with sodium nitriloacetate (NTA), which also effectively binds water hardness ions and functions as a builder. However, by the end of the same year, NTA had already passed out of favor as a result of the finding that NTA also binds (sequesters) other potentially toxic metals (e.g., cadmium [Cd] and mercury [Hg]) and could have easily released these ions in a location where the results would be quite serious (e.g., across the placental barrier to a developing human fetus). In addition, NTA contains nitrogen, which is also a good fertilizer and nutrient source for algae.

The detergent industry has since offered a variety of phosphate replacements, which prominently include other inorganic builders such as sodium carbonate, complex aluminosilicates called zeolites, and sodium silicates. As a precipitating builder, sodium carbonate (Na_2CO_3), or washing soda, acts as a builder by precipitating calcium ions, thus increasing the detergency of the surfactant system by softening the water. However, the calcium carbonate precipitate ($CaCO_3$) can elicit deleterious effects on automatic washing machines. In addition, an excess of carbonate ions within a body of water leads to a highly alkaline (basic; excess of OH^- ions) condition, which can cause washed fabrics to be irritating to the skin and eyes. Zeolites are a complex of aluminum, silicon, and oxygen. When added to hard water as part of a synthetic detergent formulation, the sodium salts of zeolites act like ion-exchange resins, exchanging their sodium ions for ions within hard water (e.g., calcium). As ion-exchange builders, zeolites soften water and hold hard water ions in suspension (rather than being precipitated), and zeolite solutions are not nearly as alkaline as sodium carbonate solutions. Sodium silicates, also known as "water glass" or soluble silicates (e.g., Na_2SiO_3, $Na_2Si_2O_5$, Na_4SiO_4), are produced from sand and sodium carbonate. As a precipitating builder, they precipitate both magnesium and calcium ions within hard water and also act as a corrosion inhibitor by protecting washing machine die-cast internal frames from rust formation during wash agitation. This builder also increases the effectiveness of the physical detergent capacity of powder detergent formulations.

An example of an organic builder included in synthetic detergent formulations is the sodium salt of carboxymethylcellulose (CMC). Although a French patent for the use of CMC as an additive to washing materials was thought to have been applied for in 1936, this patent was not developed extensively until World War II, when CMC was used in Germany. Initially used on a moderately large scale as an extender for soap, which was in short supply, it was later used as an additive to the synthetic detergents being produced as a substitute for soaps during wartime. After World War II, international intelligence reports on German industrial efforts were published, and the use of CMC as an additive to synthetic detergent powders, which eliminated redeposition of soil problems, was noted. Treating pure cellulose with caustic soda and chloracetic acid produces CMC. Also considered a suspension agent, CMC increases the negative charge in fabrics, thus causing fabrics to repel dirt and prevent redeposition of dirt particles on clothing. CMC is best suited on fabrics derived from cellulose (e.g., cotton, rayon, etc.) and on fabric blends with cellulose components.

Bleaches

In general, all bleaches are oxidizing agents, which are used to whiten and brighten fabrics and assist in the removal of challenging stains. Some types of fabric stains are bound so tightly in place that they cannot be easily dissolved and must be destroyed instead. The colors of such difficult stains are often associated with weakly bound electrons, such as those involved in double bonds between atoms. Double bonds can give organic molecules their colors, and groups of atoms that give rise to color in molecules are called chromophores. Bleaches act by attacking the vulnerable electrons of stain molecules by using electron-removing atoms, including oxygen and chlorine, and destroying the chromophores in the stain molecules. Thus, bleaches tend to convert soils into colorless, invisible soluble particles, which may then be removed by detergents and flushed out in the waste washing water. Alternately, the molecules may remain on the fabric but be no longer capable of absorbing visible light. Thus, while the oxidized form of stains is less highly colored compared with the reduced form, the oxidized form of the stain may remain on the garment in some cases.

The familiar domestic liquid laundry bleaches added separately to stained clothes are all generally aqueous (water-based) 5.25 percent sodium hypochlorite (NaOCl) solutions. This bleach is made by dissolving chlorine gas in a dilute solution of sodium hydroxide (12 to 16 percent) until the alkalinity is neutralized. The resulting solution is diluted to approximately 5 percent. This chemical acts as a bleach because the hypochlorite ion is a modest oxidizing agent that can oxidize many of the chemicals responsible for staining. Hypochlorite bleaches release chlorine rapidly, and these high concentrations of chlorine can also disinfect and deodorize fabrics. However, such high concentrations of chlorine can be quite damaging to fabrics, breaking apart fabric molecules and weakening the clothing item. These bleaches do not work well on polyester fabrics, often leading to a yellowing rather than a whitening effect. They sometimes destroy the chromophores in dye molecules, turning colored fabrics white. Other times, such bleaches modify the dye molecules and change the color of the fabric altogether. Other types of bleaches are available in solid forms, which are developed to release chlorine slowly into water to minimize the damaging effects to clothing. Symclosene ($N_3O_3C_3Cl_3$), a cyanurate-type bleach, is an example of a bleach available in solid form.

In the 1950s, a prominent innovation and addition to synthetic detergents was detergent with oxygen bleach. Oxygen-containing (color-safe) bleaches assist in removing stains from almost all types of washable fabrics and work more gently than chlorine-based bleaches. Such bleaches frequently contain oxygen compounds such as sodium perborate ($NaBO_2$-H_2O_2). As indicated by the formula, this bleach is a complex of $NaBO_2$ and the powerful oxidizing agent hydrogen peroxide (H_2O_2). When added to hot water (above sixty-five degrees centigrade), this substance decomposes into hydrogen peroxide and sodium borate ($Na_2B_4O_7$). The liberated hydrogen peroxide acts as a bleach, attacking double bonds of stain chromophores as it in turn decomposes to release oxygen (O_2). This type of oxidation removes much of the stains on clothing while generally not affecting fast (permanent) fabric coloring. The tetraborate ion formed is also useful as a laundry aid builder, as it readily forms a complex with iron (Fe). Sodium tetraborate ($Na_2B_4O_7 \cdot 10H_2O$) is also known as borax, and this product alone is marketed as a laundry aid. However, borates are somewhat toxic. While the potentially deleterious effects of borax exposure in humans are currently being investigated, boron is specifically toxic to citrus crops, and detergent runoff must be avoided during irrigation.

Oxygen-releasing bleaches are generally less active than chlorine bleaches and require higher temperatures, higher alkalinity, and higher concentrations for working efficiency. They are particularly effective when used during a clothing presoaking routine, as they require fairly high temperatures for effectiveness during the wash cycle. At lower water temperatures, an enzyme found in many biological stains (e.g., blood, grass, etc.) called catalase decomposes hydrogen peroxide quite rapidly and can destroy the perborate bleach within a few minutes. The higher water temperatures are required for oxygen-releasing bleaches to perform effectively, as the catalase enzyme is easily deactivated at high temperatures. Used mainly for bleaching white-colored resin-treated polyester/cotton fabrics, oxygen-releasing bleaches can protect these types of fabrics longer and allow them to become whiter than can chlorine bleaching techniques. The rapid increase in the use of synthetic fibers for clothing manufacture, which are adversely affected by chlorine, has increased the use of sodium perborate.

Enzymes

Another class of laundry aid substances added to laundry detergents is the enzymes, or organic catalysts. The enzymes used are very stable and are readily isolated from microorganisms. In general, enzymes are proteins that catalyze specific reactions by binding to specific chemical products called substrates. These organic (carbon-based) biological catalysts tend to hasten chemical reactions without themselves becoming altered or destroyed. In general, laundry aid enzymes act by breaking down various products (e.g., proteins and fats) that may bind stains to clothing. They act to cut up large stain molecules into smaller fragments that can then be washed away with water.

The use of enzymes for washing has a long history, starting with a patent in 1913 for soda plus a small amount of impure proteolytic (protein-digesting) trypsin enzyme marketed as a prewash. However, it had limited commercial success and was eventually replaced by a bacterial protease, which required neutral pH water conditions. Proteolytic enzymes had been tried as additives to washing powders in Germany in the 1920s and again in Switzerland in the 1930s with limited success. Eventually, better strains of enzymes were developed, with stability to a wider pH spectrum, stability against the addition of bleaches, and quicker action. In the late 1960s, a few presoaks and synthetic detergents appeared on the U.S. consumer market containing enzymes. Enzyme presoaks are used to soak items before washing to remove and decompose difficult protein-based stains and soils, including food, grass, and blood. When added to the wash cycle of automatic washing machines, they increase cleaning power. However, most enzymes are denatured at high temperatures, so detergents using enzymes usually perform best in warm (not necessarily hot) water. Technological advances are being developed (e.g., manipulation of enzymes isolated from bacteria inhabiting natural hot springs) that allow enzymes added to detergents to perform in high-temperature cleaning cycles. While enzymatic powders currently hold a large proportion of the household detergent market, the future of enzymatic powder production remains obscure, because the effectiveness of these products has been questioned and concern over their safety with repeated use has been suggested.

The enzymes used are generally proteolytic (protein cleaving/digesting) and lipolytic (fat cleaving/digesting). Protease enzymes cause proteins to be hydrolyzed back into their constituent amino acid building blocks, and lipases cause ester linkages in fats to hydrolyze. The most common protein enzyme type is alkaline protease, which digests protein in alkaline conditions. Although enzymes can digest proteins in stains, they can also cause severe allergic reactions in humans. As such, they are often used in a granulated form, coated with chemicals including polyethylene glycol, which melts and releases the enzymes within the wash cycle. Lipases have been developed to increase the cleaning of fatty soils within clothing. Some genetically engineered lipases efficiently convert fats in food, cosmetics, and sweat into fatty acids and glycerol during the spin cycle of automatic laundering and during prespotting treatment regimens. Other enzymes include amylases, which are used in detergents to degrade starches to water-soluble sugars. They perform well in cold water temperatures because they hydrolyze the "starch glue" that binds the soil to the fabric. In addition, cellulases remove cellulose-based fine, fluffy microfibrils released from cotton after repeated washings. Such microfibrils cause characteristics such as stiffness and graying coloration in some fabrics, particularly noticeable in laundered towels.

Fabric Softeners

Another class of laundry aid substances added to laundry detergents is the fabric softeners. First developed in the 1950s as products added to the final rinse cycle, fabric softeners make fabrics softer and fluffed up and decrease static cling, wrinkling, and drying time. These products remain on fabrics after laundering, allow for easier ironing of clothing, and provide a pleasant fragrance to fabrics. By the 1970s, fabric softeners were available as products added to the washing cycle, as part of multifunctional laundry products (detergents with added fabric softeners), and as separate sheets added to the dryer.

Liquid fabric softeners added to the rinse cycle are predominantly cationic surfactants. Cationic surfactants consist of two parts: a long water-insoluble hydrocarbon tail region and a small positively charged water-soluble head region. The most common cationic surfactants are called quaternary ammonium salts, because they possess four hydrocarbon groups attached to a nitrogen atom bearing a positive charge. These compounds are based on a positive ammonium ion, which itself is based on a positive nitrogen ion. In fabric softeners, a positive centrally located nitrogen ion forms covalent bonds with four hydrocarbon chains. A specific type of quaternary ammonium salt, which bears only two long carbon chains and two smaller groups of nitrogen and is used as a fabric softener, is called dioctadecyldimethylammonium chloride. The long hydrocarbon chains are hydrophobic and have a physical characteristic similar to most oily lubricants. Normally, when a fabric softener is applied to wet negatively charged fabric fibers, the softener will stick to the fabrics very strongly. Cationic surfactants are seldom used within the actual wash cycle, as washing detergent surfactants are predominantly anionic, possessing a negatively charged head region. The oppositely charged head regions would thus tend to clump together and precipitate from solution, destroying the detergent action of both.

As stated above, the cationic (positive) charge of fabric softeners has a strong affinity for wet negatively charged fabrics. As such, fabric softeners form a uniform layer one molecule thick on the surface of clothing fibers. The long hydrocarbon chains lubricate the fibers and reduce friction and static. Previously treated clothing items are thus lubricated and experience weaker frictional forces when transferred to the automatic dryer. They also transfer fewer electrical charges as they tumble dry. As such, clothing flexibility and softness are both increased. However, while fabrics then possess a fluffy appearance and feel, the hydrophobic coating of the softener slightly reduces the ability of the fabric to absorb water. This is an issue for items such as towels and cotton baby diapers. In addition, cationic surfactants, although not good detergents, are mildly antiseptic, imparting a germicidal effect. They act as bactericidal agents because they coat, smother, and kill bacteria, and they may inactivate and cause imbalances in bacterial metabolic pathways and enzyme cascades.

Optical Whiteners (Fluorescers)

White fabrics, such as cotton, tend to absorb an increased amount of blue light and begin to appear yellow. Historically, "washing blue" was added when washing clothes so that cotton fabrics naturally aging to a yellowish hue would appear white. This blue dye absorbed red and green light, balancing the blue absorption of the fabric itself so that the fabric appeared colorless. Instead of using bluing, optical whiteners, or brighteners, are fluorescent dyes often added to modern synthetic detergents to increase the brightness of white fabrics. These brighteners are actually fluorescent dyes, organic compounds called blancophors (or colorless dyes; e.g., blancophor R), which contain four five-carbon rings and two centrally located $SO_3^-Na^+$ groups that increase the solubility of the compound. They coat the fabrics during washing and convert the invisible ultraviolet (UV) light/radiation component of sunlight into an almost imperceptible blue tint. Instead of absorbing red and green light to balance the white appearance of fabrics, optical whiteners reintroduce the missing blue light. The absorption of this invisible light (mostly UV-A) is reemitted as visible light at the blue and violet part/end of the UV light spectrum. These brighteners restore the mixture of colors reflected to what a white fabric would naturally reflect. When exposed to sunlight, fabrics appear brighter, almost "whiter than white," and the blue light camouflages any fabric yellowing. Many new clothing items already contain such dyes, but repeated washings remove them from the fabric.

Optical whiteners do not perform any cleaning action; thus, their only benefit is cosmetic, yielding more pleasing laundry. While clothing in fact may actually be dirty, optical whitening provides the deception of cleanliness. Because different types of fabrics carry different electrical charges (e.g., nylon has a positive charge, cotton a negative charge), oppositely charged fluorescers are often needed.

AUTOMATIC DISHWASHING DETERGENTS

First developed in the 1950s, automatic (machine) dishwashing detergents, in addition to removing soils originating from food particles and holding the particles in suspension, sequester minerals that contribute to water hardness, emulsify greasy oils, and assist in allowing water to rinse off dishes via a sheeting action. Generally, these detergents are quite caustic, and their efficiency can depend on their physical characteristics (e.g., granular powder). A typical powder-based formulation contains 30 percent anhydrous sodium tripolyphosphate ($Na_5P_3O_{10}$), 30 percent anhydrous sodium metasilicate (Na_2SiO_3), 37.5 percent anhydrous sodium sulfate (Na_2SO_4), a 2 percent chlorine bleaching chemical (e.g., sodium dichloroisocyanurate with 56 to 64 percent available chlorine), 0.5 percent low-foaming surfactant, and up to 9.5 percent corrosion inhibitors (e.g., aluminum salts). In addition, perborates, clay, perfumes, and color are often added. While chlorine-containing powders may lead to the deterioration of plastic kitchenware, powders without chlorine fail to remove the tannin stains from such beverages as teas. Corrosion inhibitors are added to protect aluminum components (e.g., aluminum saucepans). Machine dishwashing detergents generally contain a maximum of 2 percent surfactant, which consists of a low-foaming, nonionic surfactant (e.g., block copolymers, or propylene oxide and ethylene oxide). Nonionic surfactants are low sudsing and do not ionize in solution; thus, they have no electrical charge. They are also resistant to the chemical effects of hard water and clean well on most soils. For this reason, automatic dishwashing detergents produce little to no foaming action. Individual foam films tend to take up and hold particles of soil that have been removed from an item, preventing the soil particles from redepositing on cleaned surfaces. However, if foamy suds were to form within a dishwashing machine in appreciable quantities, rinsing/removal of the foam would cause serious problems and interfere with the washing action of the machine. Automatic dishwashing detergents also suppress natural foam accumulation caused by protein-based soils. Therefore, these detergents are designed not to foam and rely mainly on their strong alkalis and the vigorous action of the machine hardware for cleaning.

Examples of the alkalis included in dishwashing detergent formulations are the carbonate, silicate, or phosphate salts of sodium or potassium (examples noted above), alone or in combination. However, preparations with greater than 5 percent sodium carbonate tend to cause sheet erosion of glassware, which gradually thins glasses and leads to breakage. All of these compounds readily dissolve in water to yield a very high (alkaline/basic) pH solution. This alkaline environment at high temperatures easily dissolves and disperses greasy food residues. Although dishes can withstand such basic conditions, human skin (i.e., hands) cannot, so automatic dishwashing detergents are not recommended for use as hand dishwashing detergents. Sodium metasilicate, in particular, is quite caustic and very dangerous if swallowed. In fact, it is children who tend to be poisoned after ingestion of automatic dishwashing detergents from powder-in-door receptacles on the washing machine or directly from the box dispenser. After ingestion, alkalis within the detergent interact with fatty throat tissues and change the fatty tissues into soap, a process called "liquefaction necrosis." The nature and severity of an associated injury depends on the concentration ingested, pH, quantity, physical form (liquid or powder), and duration of exposure.

Liquid formulations have since been introduced as well and often contain a bentonite-based clay substance to prevent the liquid product from seeping out of a standard automatic dishwashing machine powder dispenser.

2

Cosmetics and Bathroom Products

BLUSH MAKEUP

Ancient Greek and Roman women were known to apply naturally occurring red pigments such as ocher, fucus, cinnabar, henna, safflower red, or cochineal as a type of rouge (red coloration) on their cheeks. Early European women used Brazilian redwood directly after access was gained to Western hemisphere resources, and red lead was also used until the early 1920s. Although early rouges were ointments, modern manufacturing allowed the blusher to be marketed in many forms, including liquid suspensions, emulsified creams and lotions, water-free creams, and hydrous and anhydrous gels. Currently, the pressed powder is the blusher marketing style choice because of its desirable matte finish after application and long-lasting wear potential.

Major components of the pressed powder blusher include the powder phase and binder (or oil) phase. Components of the powder phase include mineral powder fillers such as talc (a magnesium silicate), mica (a magnesium aluminum silicate), sericite (a form of hydrated mica), and kaolin (known as China clay). Talc is the most popular of all fillers used, as it tends to be virtually transparent depending on the particle size, and it is very soft to the touch. Although mica is also used extensively as a result of its transparency and smooth texture, it often exhibits a shiny appearance with skin application and has poor compression characteristics when used in a pressed powder. Modern non-oil control formulations rarely contain kaolin, as it tends to exhibit course texture, an extremely matte appearance, and excessive oil-absorption capabilities. Dry binders are also used to allow the compressed powder to retain form; these include metallic soaps such as zinc stearate and magnesium stearate and polymeric materials such as polyethylene. Also included in the dry powder phase of the blusher are colorants, including carmine, titanium dioxide, iron oxides, chromium oxide greens, ultramarines, manganese violet, and yellow and red lake colors. Other types of chemicals included in blushers are spherical materials that provide a smooth texture for the product and allow for the optical effect of "soft focus" after application, such as silica, nylon, and polymethyl methacrylate. The oil phase or binder of the pressed powder blusher product consists typically of oils (e.g., sunflower oil, coconut oil, castor oil, mineral oil), esters (e.g., sorbitan ester), and/or waxes that provide a creamy texture (e.g., beeswax, candelilla wax, lanolin, carnauba wax). In addition, ingredients such as vitamins (e.g., tocopheryl acetate, tocopherol [vitamin E], retinyl palmitate, ascorbyl palmitate, panthenol), herbal extracts (e.g., comfrey root, rosemary, *Aloe barbadensis*), and preservatives (e.g., methylparaben, propylparaben, butylparaben, imidazolidinyl urea) may also be added.

BODY LOTIONS AND CREAMS

During the First Dynasty of Egypt (approximately 3100–2907 BC), men and women used perfumed oils (stored in unguent jars made of alabaster and marble) to maintain soft, supple, and unwrinkled skin. By the middle of the first century AD, Romans were recommended to use a face pack of barley-bean flour, egg, and mashed narcissus bulbs to promote smooth skin. A Greek physician of the second century AD is thought to have invented cold cream containing water, beeswax, and olive oil. When rubbed on the face, the water evaporated, cooling the skin. (Cold cream formulations of the 1920s were of similar composition.) By 1700, smooth skin was in fashion and women applied oiled cloths on their foreheads and wore gloves in bed to prevent wrinkles. Herbal lotions were often used by Europeans to improve complexions scarred by various diseases, and the twentieth century led to the major industrial development and marketing of facial beauty lotions.

Creams and lotions are applied to the skin, a complex organ responsible for covering and protecting the body from damaging foreign matter and pathogens, regulating body temperature, producing vitamin D, and sensing external environmental stimuli. The skin itself is composed of two major layers, the underlying dermis that supports the upper layer called the epidermis. The dermis contains blood vessels, nerves, sweat glands, specialized receptors, and the active portion of hair follicles. The epidermis consists of several tiered layers of cells that are continuously dividing as they move toward the outside of the skin, until they lose their ability to divide and die as they reach the outermost layer. The lifeless outermost layer is called the stratum corneum, the horny ten-micrometer layer consisting of twenty-five to thirty tiers of dead cells with 10 percent water content containing mainly the fibrous protein keratin. Sebum, a fat/wax/free fatty acid oily secretion of the sebaceous glands located at the hair follicles, protects the skin naturally from loss of moisture, lubricating and softening dead keratin, thereby lowering the rate at which water evaporates from the skin surface. The skin may be further protected, and skin dryness prevented or relieved, by the application of cosmetic emollient (skin-softening) lotions or creams that act similarly to sebum. These lotion and cream colloidal dispersions consist of two or more liquids insoluble in each other (e.g., oil dispersed in water [lotion] or water dispersed in oil [cream] emulsions). The essential ingredient of each is a fatty or oily substance that forms a protective film over the skin, thereby retaining skin moisture and secondarily skin flexibility. A typical cold cream formulation was historically an emulsion of approximately 55 percent mineral oil, 19 percent rose water, 13 percent spermaceti (wax derived from sperm whales), 12 percent beeswax, and 1 percent borax (mineral including sodium borate and water).

Typical ingredients that are currently used to enhance skin barrier functioning include mixtures of alkanes derived from petroleum (e.g., mineral oil or petroleum jelly) and dimethicone. Ingredients that provide sustenance to the dermal lipid barrier include natural fats (e.g., lanolin), collagen (protein in connective tissue), and oils (e.g., olive oil, sweet almond oil, coconut oil, apricot kernel oil, sunflower seed oil, macadamia nut oil, orange oil, and corn oil). Other ingredients added include water, fragrances, pigment colorants, waxes (e.g., beeswax, orange wax), emulsifiers (e.g., cetyl alcohol, cetearyl alcohol, polysorbate 60, propylene glycol, and glycerin), opacifying agents and thickeners (e.g., glyceryl stearate, magnesium aluminum silicate, and xanthan gum), vitamins (e.g., tocopheryl acetate and tocopherol [vitamin E]), herbal extracts (e.g., rosemary, *Aloe barbadensis*, and matricaria), and preservatives (e.g., disodium EDTA, diazolidinyl urea, citric acid, methylparaben, propylparaben, ethylparaben, butylparaben, and butylated hydroxytoluene [BHT]).

Cleansing creams and lotions are detergent-based or emulsified oil systems that are designed primarily for the removal of surface oil, pollutants, or cellular debris along with makeup from the face and neck areas. Most emulsified cleansing creams are manufactured similar to cold creams but modified to enhance their debris-removal capability. They usually contain from 15 to 50 percent oils (e.g., mineral oil, vegetable oils, fatty esters, and propoxylated oils) with limited quantities of waxy materials.

Antiwrinkle Creams

While even brief exposure to certain wavelengths of UV radiation may elicit negative changes in the skin, including wrinkles and sagging, through the destruction of the skin-firming protein collagen, specialized facial emulsions frequently contain sunscreen protection (e.g., dioxybenzone) along with additional active ingredients designed to normalize the facial and neck skin and around the eyes. Designed to improve the facial skin condition with prolonged daily and/or nighttime use, the active ingredients in these products are derived from a variety of plant sources. Formulas containing emulsifiers (e.g., lanolin, cyclomethicone, or dimethicone) along with vitamin A derivatives, α-hydroxy acids, various fruit acids (e.g., lactic, glycolic, malic, and citric), yeast extract, and butylated hydroxyanisole (BHA) derivatives (e.g., willow bark extract) are designed to rebuild, resurface, repair, and diminish the appearance of fine facial lines while protecting the face from exposure to detrimental environmental factors. A typical vanishing cream, which smooths wrinkles so that they appear to fade away, is often composed of 70 percent water, 20 percent stearic acid (saturated fat), and 10 percent glycerin, with small amounts of potassium hydroxide, preservatives, and fragrance.

DEODORANT/ANTIPERSPIRANT

Sweat (perspiration) is the fluid produced by the secretory portion of the 3 to 4 million sweat (also called sudoriferous) glands located in the subcutaneous and/or dermis of the skin. There are two types of sweat glands, eccrine and apocrine, differentiated by their structure, location, and secretion chemistry. Eccrine glands are distributed throughout most of the skin and are highly concentrated on the palms, soles, forehead, face, and axillae (armpits). These glands function throughout life and secrete a slightly acidic solution containing a mixture of water, inorganic ions (e.g., sodium, potassium, and chloride), lactic acid, ascorbic acid, amino acids, urea, uric acid, ammonia, and glucose. This secretion tends to have no odor, and the body may be generally cooled after the evaporation of skin surface water from the eccrine-based sweat. However, bacteria (e.g., *Staphylococcus* species) may act on perspiration residue and sebum (natural body oil) to produce unpleasant odiferous products such as short-chain fatty acids and amines.

Apocrine glands, which secrete their products into hair follicles, are concentrated mainly in the skin of the axillae, pubic region, and pigmented areas of the breasts. Apocrine glands first function at the start of puberty and produce a more viscous solution than eccrine glands. Apocrine glands are typically stimulated during emotional stresses and sexual excitement. While these apocrine secretions alone yield little odor, bacteria that colonize nearby hair follicles may degrade the contents of apocrine fluids, producing malodorous chemicals.

Deodorants are products that typically contain a fragrance (e.g., musk scent from ethylene dodecanedioate) or perfume to disguise offending body odor and a germicide to destroy odor-producing bacteria. The germicide is usually a long-chain quaternary ammonium salt, various salts of zinc, a phenol such as triclosan, a chemical such as benzethonium chloride, or a broad-spectrum antibiotic (e.g., neomycin). In 1888, an unknown inventor from Philadelphia formulated deodorant, which contained zinc oxide. It was generally recognized as the first product to prevent odor and kill bacteria. Deodorants subsequently developed in the early twentieth century contained noxious chemicals such as cresylic acid or hexachlorophene.

Antiperspirants are usually deodorants that additionally impede the production of perspiration. Most antiperspirants contain active ingredients such as the aluminum chlorohydrates [e.g., $Al_2(OH)_4Cl_2$ and $Al_2(OH)_5Cl$] or aluminum zirconium tetrachlorohydrex glycine complex. Aluminum salts function as astringents, producing an insoluble hydroxide gel in the sweat pores and thus physically blocking the release of secretory products by constricting the opening of the sweat gland ducts. The active ingredients in antiperspirants may also reduce odor, most likely by destroying bacteria that decompose the organic portion of the secretions. Because sweating is considered a natural, healthy body process, regular bathing and changing/washing of clothes make antiperspirants unnecessary for most individuals.

Deodorants and antiperspirants can be formulated into creams or lotions typically containing an oil base (e.g., cyclomethicone), solvents (e.g., polypropylene glycol [PPG] 14 butyl ether, stearyl alcohol, glycerin, or polyethylene glycol [PEG] 8 distearate), emollients (e.g., cyclopentasiloxane or dimethicone), lubricants (e.g., hydrogenated castor oil or dimethicone copolyol), humectants (propylene glycol or dipropylene glycol), and antioxidant/preservatives (e.g., BHT, T-butyl hydroquinone, or citric acid). Specialized brands may contain silica, talc, opacifying agents (e.g., glyceryl oleate), or antiwhitening agents (e.g., phenyl trimethicone). Deodorants and antiperspirants can also be dissolved in a solvent (e.g., denatured [ethanol] [SD] alcohol 40 or propylene carbonate) and applied as an aerosol with the use of a nonchlorofluorocarbon propellant (e.g., isobutane, butane, propane, 1,1-difluoroethane, or hydrofluorocarbon 152a). The first antiperspirant aerosol deodorant was introduced in 1965. Modern aerosol antiperspirant deodorants also may contain an oil base (e.g., cyclomethicone), emollients (e.g., dimethicone), talc, silica or a modified magnesium aluminum silicate (e.g., quaternium-18 hectorite), emulsifiers (e.g., isopropyl myristate), and fragrance.

EYE SHADOW

Ancient Egyptian women developed the art of decorating the eyes by applying dark green color to the lower lid and blackening the upper lid with kohl, a preparation composed of antimony and soot. Egyptians also used powdered charcoal, powdered galena, or soot alone to achieve the same eyelid-darkening effect. Semiprecious stones such as lapis lazuli and malachite were also ground and used as eye shadows. In addition, the Romans adopted the technique of using kohl to darken the eyelids. In the Middle Ages, the Crusaders found eyelid-coloring cosmetics widely used in the Middle East, and it was they who spread the use of these types of cosmetics throughout Europe. Elizabethan women used an iridescent eye shadow made of ground mother of pearl. By the nineteenth century, research in France led to the development of more and better eye cosmetics at decreased cost.

Eye shadow is a cosmetic product applied to the eyelids for coloring. Modern eye shadow is mostly composed of a petroleum jelly base with fats (e.g., mineral oil or jojoba oil) and waxes (e.g., beeswax, lanolin, or ozokerite). It is colored with dyes that include ultramarine colors (organic polymers containing aluminum, oxygen, silicon, sodium, and sulfur) such as blue, pink, or violet, iron oxides of various shades, carbon black (a form of carbon resembling charcoal), carmine, bismuth oxychloride, manganese violet, chromium hydroxide greens, bronze powder, aluminum powder, ferric ferrocyanide, ferric ammonium ferrocyanide, zinc oxide (ZnO), or titanium dioxide (TiO_2). As titanium dioxide is a white opaque powder, it may serve as a base, allowing other included colored dyes to be viewed by muting the natural color of the skin. Eye shadow may also contain additional chemicals to enable a longer shelf-life and manufacturing consistency for the product, including talc, aloe, binders (e.g., octyl palmitate), and preservatives (e.g., parabens, imidazolidinyl urea, or BHA). A typical composition may be approximately 60 percent petroleum jelly, 10 percent fats and waxes, 6 percent lanolin (grease in sheep's wool), and the remainder dyes, pigments, and preservatives. Unfortunately, the addition of synthetic chemical preservatives is often the source of consumer allergic reactions, including contact dermatitis.

FACE MASKS

The earliest known recordings of skin-treating facial mask use date back to the Egyptians, and their facial mask formulas were known to be pastes including honey, milk, and vegetable flours. Romans used masks formulated with ass's milk, wet bread dough, or crude wool grease combined with honey, eggs, barley flour, crushed beans, narcissus bulbs, orris root, powdered horns of cows, and seabird excrement. Throughout medieval times, Asian women used facial masks made of pearl powder, white jade, ginseng, lotus seed powder, and camphor. In the late nineteenth and early twentieth centuries, European women mixed yeast with water or used yogurt alone as a cosmetic facial mask. By the mid 1960s, the back-to-nature movement led to the use of ingredients such as eggs, honey, milk, cream, strawberries, cucumbers, olive oil, oatmeal, and mayonnaise in "homemade" mask recipes. At the start of the twenty-first century, modern commercially developed facial masks include natural product ingredients, along with chemically advanced active ingredients, to support the competitive consumer-driven cosmetic market.

Facial masks are developed to target consumers with different skin types, classified as dry, normal, or oily. Specialty products targeting consumers with special skin needs include those with combination, acne-prone, sensitive/allergy-prone, stressed, or aged skin. Depending on the active ingredients and the chemical base formula, facial masks are characterized by functionality and include clay masks, peel-off masks, cream (hydrating) masks, and exfoliating masks.

Clay masks are applied as an even film over clean skin and draw (via capillary action) into them materials that are absorbable or adsorbable, thereby acting as a mild abrasive and exfoliant. Clay masks, generally targeted for normal, oily, or acne-prone skin types, are based on fine particle-sized or micronized solids such as adsorptive clays, including bentonite, hectorite, magnesium aluminum silicate, kaolin, green, red, and pink clays, magnesium carbonate or oxide, zinc oxide, lake- or river-based silts, titanium dioxide, or colloidal oatmeal. Clays in general are derived from silico-aluminum sedimentary rocks, and their coloration is dependent on the types of trace metals present. Other ingredients included in clay mask formulations include water, opacifying agents (e.g., titanium dioxide, zinc oxide), thickeners (e.g., methylcellulose, ethylcellulose, carboxymethylcellulose, polyvinyl pyrrolidine [PVP] and PVP/PVA [polyvinyl alcohol] resins, xanthan gum, carbomers, polyacrylates, sodium alginate, acacia), emollients (e.g., glycerin, propylene glycol, allantoin, mineral oil), emulsifiers (e.g., disodium cocamido monoisopropylamid [MIPA]-sulfosuccinate, isopropyl palmitate, ceteareth-5, polysorbate 60, cetyl alcohol), humectants (e.g., sorbitol), alcohol (e.g., SD alcohol 40, phenoxyethanol), preservatives (e.g., diazolindinyl urea, methylparaben, propylparaben, disodium EDTA), and fragrance.

Peel-off masks are applied in a uniform layer on the skin and upon drying and removal produce a sensation of skin tightness and cleansing action. Peel-off masks are marketed to consumers with normal, oily, combination, and acne-prone skin types. These masks are believed to perform the functions of skin exfoliation, hydration, and purification. They are based on plasticized polyvinyl alcohol. Hydrophilic emollients (e.g., ethoxylated fatty acid or alcohol derivatives, dimethicone copolyol) and humectants (e.g., glycerin, propylene glycol, sorbitol) are added to prevent moisture loss and cracking of polyvinyl alcohol film. Solvents (e.g., ethyl alcohol, SD alcohol 40), preservatives (e.g., urea, parabens), and fragrance are also often added to ensure product consistency, purity, and esthetic consumer appeal.

Cream masks are applied to clean skin, whereby the skin absorbs the cream emollients from the mask and leads to a feeling of soft and moist skin after mask removal. They are marketed to consumers with dry, tight, rough, aged, or environmentally stressed skin types. The formulations consist of heavy-textured emulsions based on water-in-oil or high-oil-content oil-in-water bases. While emollients (e.g., PPG 30 cetyl ether, glyceryl stearate, PEG 100 stearate) serve as the primary ingredients, additives including emulsifiers and humectants (e.g., cetearyl alcohol, myristyl myristate, mineral oil, petroleum, propylene glycol, stearic acid, sorbitol) are used to provide skin moisturization, softening, lubrication, and protection. Thickeners (e.g., carbomer, honey), coloring agents (e.g., titanium dioxide, mica, iron oxides, caramel), preservatives (e.g., urea, parabens), and fragrance are also commonly included in these formulations.

Exfoliating masks are formulated to allow for the physical and/or chemical removal of nonliving cells from the upper epidermal layer of the skin that are shed through a natural process. Continuous and evenly distributed skin exfoliation may assist in preventing dry skin and acne formation. Acting on the basis of friction, physical (mechanical) exfoliating masks usually consist of polyethylene grains, wax beads, ground apricot or walnut shells, cornmeal, sodium chloride crystals, or encapsulates. Chemical exfoliating masks penetrate the upper epidermal skin layer and loosen bonds that hold dead shedding cells to the skin surface, thereby helping to refine skin texture and improve the appearance of wrinkled, age-marked, and UV-damaged skin. Active ingredients included in these formulations include enzymes and α- and β-hydroxy acids, often extracted from milk, fruit, sugarcane, willow bark, wintergreen leaf, or sweet birch bark sources. Other additives included in these formulations are water, emollients (e.g., mineral oil, lanolin, glycerin, butylene glycol, petroleum, wax, dimethicone, shea butter), humectants (e.g., sorbitol), emulsifiers (e.g., stearic acid, glyceryl stearate, ceteareth-20, cetyl alcohol), surfactants (e.g., sodium lauryl sulfate), clays (e.g., magnesium aluminum silicate, talc, kaolin), thickeners (e.g., carbomers, algae), antioxidants and preservatives (e.g., glycolic acid, citric acid, lactic acid, parabens), and fragrances.

Special additives frequently included in all facial mask formulations are different types of fruit and herbal extracts and oils (e.g., avocado, grape, orange, lemon, tangerine, lime, grapefruit, cucumber, macadamia nut, green tea, oat, rice bran, matricaria, primrose, jojoba, *Aloe vera*, witch hazel, chamomile, peppermint, sage, horsetail, comfrey, lavender, rosemary, geranium, sunflower, rose hip seed, safflower, cornflower, dandelion, fennel, milk thistle, clove, Saint John's wort, ginseng, sea minerals), along with vitamins (e.g., panthenol, tocopheryl acetate, ascorbic acid, retinyl palmitate, cholecalciferol) that are used to target specific skin conditions or problems.

FACE POWDER

Ancient Greek and Roman women developed the beauty treatment of using white lead and mercury on faces to achieve a chalky complexion. This was a dangerous practice, as these heavy metals were subsequently absorbed through the skin and resulted in many deaths. While early physicians recognized this practice as a problem, the European Middle Ages followed the Greco-Roman trend of achieving smooth complexions and pale whitened faces with the use of lead (usually a composite of carbonate, hydroxide, and lead oxide). To fight the destructive effects of lead applied on the face, masks were synthesized using ground asparagus root and goat's milk. Fashionable sixth-century noblewomen would often achieve a pale look by bleeding themselves. Women of the Italian Renaissance also continued the use of destructive lead paint on their faces, necks, and cleavage. In Elizabethan England, women still used white lead face paint, and the lead was mixed with vinegar to form a paste called ceruse. The white lead made hair fall out, and the extensive use of ceruse throughout the Elizabethan era is linked to the presence of the fashionable high foreheads as hairlines receded. The use of lead-based ceruse on complexions continued throughout the reign of Charles I. While Victorians avoided the use of almost all makeup, at the turn of the nineteenth century, when makeup regained acceptance, zinc oxide was discovered to make a good face powder that did not harm the skin.

Modern face powders are designed to provide a smooth complexion, diminishing the glossy shine that results from the accumulation of sebum and perspiration. They also may add an agreeable coloration, texture, and odor to the skin without drawing attention to their own presence as a makeup. The properties of a good face powder include covering power, adhesiveness, absorption, matte effect, and slip (application smoothness lacking drag). The bulk (by percentage of weight) primary texturing consistency of a face powder is usually attributable to components such as talc (magnesium silicate), kaolin (clay), or mica (silicate minerals). Secondary components added may include materials such as zinc oxide (zincite mineral), calcium carbonate (chalk; limestone salt), titanium dioxide (white opaque powder), or cornstarch. These secondary components may provide additional skin protection, cover, and oil-absorption properties to the product. Compact or pressed face powders are wet by binders and emollients that also increase application ease and might include mineral oil, nylon-12 (petrochemical), dimethicone (silicone), glycerin, lauroyl lysine, or phenyl trimethicone. In addition, lubricants (e.g., octyldodecanol stearoyl stearate), texturizers (e.g., magnesium stearate, zinc stearate), binders (e.g., calcium silicate, cetyl alcohol), softeners (e.g., lanolin), preservatives (e.g., methylparaben, propylparaben, imidazolidinyl urea), antioxidant moisturizers (e.g., retinyl [vitamin A] palmitate, tocopheryl [vitamin E] acetate, ascorbyl palmitate), texture enhancers (e.g., silica), and color pigments (e.g., iron oxides, ultramarines, D&C and FD&C Lake colors, green oxides and hydroxides, carmine) are often added. Specialized product characteristics are achieved with the addition of chemicals such as boron nitride (white graphite, which provides a silky luxurious feel with increased product wear and spreadability) or bismuth oxychloride (skin protective salt, which allows for a frosty "pearlescent" product appearance).

HAIR COLORING

Available in a variety of colors and shades, hair-coloring vegetable extracts and powdered metals have been used since the earliest record of human history as cosmetic agents. While Greek women were known to dye their hair black, Queen Elizabeth I dyed her hair red. Henna, a vegetable dye with an active component called lawsone (2-hydroxy-1,4-naphthoquinone), has been used throughout the centuries to give hair a red-orange tint. In addition, weak solutions of hydrogen peroxide were and currently are used as hair bleaches. A French chemist formulated the first synthetic hair dye in 1907. It was this basic formulation that the modern technology of changing hair color, accomplished through the reactions of complex synthetic organic and organometallic chemicals, is based upon.

Hair contains two natural pigments located within the inner hair cortex. These pigments are melanin, a brown-black pigment, and pheomelanin, a red-brown or yellow-brown iron-based pigment chemically similar to melanin. Melanins are biosynthesized in the hair follicle by a series of enzymatic reactions with the amino acid tyrosine. The color of hair (e.g., light blond, brown, blue-black, auburn red) depends on the amount and physical conditions of these two pigments within the strands (with redheads possessing a high concentration of pheomelanin). The absence of any pigment leads to the appearance of white or gray hair. Pigment-containing granules are usually present within the hair central cortex but also may be present in the surrounding outer hair cuticle. Dark hair (compared with light or blond hair) has both more granules and more pigment per granule.

Darkening the natural color of the hair can be accomplished through the action of temporary dyes that act similarly to a hair conditioner. Most temporary chemical dyes contain the salts of large acidic molecules (e.g., FD&C Blue No. 1) that do not easily penetrate the hair outer cuticle. A colored film forms over each hair and washes off with the action of a single shampooing. Semipermanent dyes consist of small molecular alkali bases such as nitrobenzene, azo, and anthraquinone dyes (e.g., 2-amino-4-nitrophenol [yellow], 4′-hydroxy-2-nitrodiphenylamine [orange], 1,4-diaminoanthraquinone [violet], 1,4,5,8-tetraaminoanthraquinone [blue]) that, when applied to the hair, soften and swell the cuticle cells and allow the small synthetic pigment molecules to move between the hair shaft and the outer cuticle. Because these pigments do not enter the hair shaft, some of the pigment molecules wash out with each shampooing event. Overall, the applied hair-darkening color gradually fades over time.

Specialized hair treatments that are designed to darken graying hair gradually are often based on the action of metal atom incorporation into organic molecules. A solution containing colorless lead acetate [$Pb(CH_3COO)_2$] is applied to the hair and penetrates the hair shaft to form brown-black lead sulfide (PbS) upon reaction with sulfur atoms within the cystine amino acids of the hair protein keratin. Repeated applications result in a gradual darkening of the hair as the concentration of lead sulfide increases.

Converting dark hair to a lighter shade requires first bleaching (oxidizing) the natural pigments in the hair cortex to colorless products, usually with an oxidizing agent solution such as hydrogen peroxide (H_2O_2), frequently followed by application of a synthetic permanent hair dye. The addition of synthetic dyes without first bleaching the hair results in hair darkening. Hydrogen peroxide, especially when used with an alkalizer such as ammonia, causes the hair cuticle scale cells to swell and separate and softens the cortex to allow the pigment granules to enter the cortex through the cuticle and hair shaft. Oxidation also prevents the synthetic dye granules from migrating out of the cortex once they have entered. A bleach "booster," such as peroxydisulfates (persulfates), is often added to the peroxide to increase its effectiveness.

Permanent hair dyes are applied as small colorless alkali molecules that, once they have entered into the hair cortex, react with the hydrogen peroxide in a redox reaction to form giant clusters of color that shampooing will not remove. The permanent hair dye pigments are commonly applied as two separate small mobile alkali organic intermediates after bleaching, a primary intermediate and a secondary intermediate (called the coupler). While the initially applied primary intermediate is oxidized by hydrogen peroxide within the cortex, the coupler combines with the product of this redox reaction to form an immobile permanent coloring effect. The products of these reactions probably include quinones and nitro compounds. Thus, darker shades may be formed as newly introduced dyes add to the melanin color already present, and lighter shades may be formed as the melanin is bleached and dyes induce lighter color formation. As permanent color dyes only affect the cortex of the hair, newly grown hair will have natural color as it emerges from the scalp.

Permanent hair dyes are often derivatives of an aromatic amine, discovered as a hair dye in 1883, called *para*-phenylenediamine (H_2N-C_6H_4-NH_2), including primary intermediates such as *para*-aminophenol, dihydroxybenzene, 4-methylaminophenol, tetraaminopyrimidine, 2-aminomethyl-4-aminophenol, and *para*-aminodiphenylamine. The *para*-phenylenediamine compound produces a black color, with its derivative *para*-aminodiphenylaminesulfonic acid used in blond formulations. Couplers are generally aromatic phenols or amines that have an available site that is subject to electrophilic interaction. Typical couplers include chemicals such as resorcinol, *meta*-phenylenediamine, 2,4-diaminoanisole, *meta*-aminophenol, 1,5-dihydroxynaphthalene, 1-naphthol, and 4-amino-2-hydroxytoluene. A variety of colors may be produced with the combination of many primary intermediates and couplers that undergo consecutive reactions to produce a mixture of dyes leading to the final shade. Most over-the-counter permanent hair-dying kits contain separate darkened containers for oxidizing and intermediate agents and include the addition of alkalizers (e.g., ammonia, ammonium hydroxide, monoethanolamine, triethanolamine), solvents (e.g., propylene glycol, water, ethanol, carbitol), stabilizers (e.g., phosphoric acid), antioxidants (e.g., ascorbic acid, sodium sulfite, BHT), metal ion sequestrants (e.g., EDTA), viscosity-enhancing chemicals (e.g., polyacrylic acid derivatives, nonoxynol-9 and -4, hydroxyethyl cellulose, oleyl diethanolamine), conditioners (e.g., polyquaternium-6, quaternium-40, distearyl dimethylammonium chloride), preservatives (e.g., parabens), and fragrance.

HAIR CONDITIONERS

The natural condition of the hair depends on the physical state of the outer protective sheath of the hair shaft (called the cuticle). The cuticle consists of dead cells arranged as stacked transparent plates that reflect light. Factors such as general environmental and chemical exposure, excess heat, and/or excessive mechanical styling dislodge and fray the tile cells, thereby allowing the hair shaft to be exposed to dehydration by progressive cuticle deterioration. The cortex becomes susceptible to damage, especially in the area close to the fiber tips. In addition, the hair takes on a dull appearance as the hair in general has reduced reflective capacity, and the raised tile plates cause the hair to be more susceptible to tangling. Conditioners help keep hair from tangling by leaving a waxy coating on the hair shaft, thereby smoothing the cuticle tile plate cells down in place. Conditioners in general contain active materials that are adsorbed or absorbed onto the hair surface, thereby changing the surface energy and friction, providing gloss, shine, enhanced tactile properties, manageability, and body to otherwise potentially dry, brittle, damaged, and dull (non-light-reflective) hair. Different types of conditioners include conditioners and cream rinses (to be applied after shampoo and rinsed off shortly thereafter), leave-in conditioners (designed to be left on hair after application with no rinsing), and deep conditioners (to provide intensive treatment; designed to be left on hair between five and twenty minutes and then rinsed off). Generally, conditioner ingredients include water, primary conditioning agents (quaternary ammonium salts, cationic polymers, silicones), refatting agents (fatty alcohols, waxes), secondary conditioning agents (natural oils, silicones), emulsifier (typically nonionic surfactants), specialty additives (proteins, resins, dyes), thickeners, pH adjusters, fragrance, and preservatives.

Hair consists mainly of the protein keratin, and in general, the sorption of a cosmetic ingredient onto hair is governed by its attraction or binding with the negatively charged composition of keratin amino acids. Other structural elements, such as lipids, may also contribute to the dynamics of colloidal structure adsorption onto hair. Modern conditioning products are the result of the development of mild detergents and cationic surfactants in the 1940s and 1950s. Modern conditioners consist mainly of long-chain alcohols or long-chain quaternary ammonium salts that coat hair fibers. Cationic surfactants such as quaternary ammonium compounds (e.g., cetrimonium chloride, stearalkonium chloride, quaternium-18, -27, and -80) are the salts of strong acid or strong base and contain a hydrophilic cationic group (attaches to keratin) carrying one or two hydrophobic hydrocarbon chains (provides conditioning). In addition, cationic polymers (e.g., acrylates copolymer, guar hydroxypropyltrimonium chloride, polyquaternium-4, -24, -11, -28, -6, -7, -39, -16, and -2) may be used as conditioning agents, as they also display high affinity for negatively charged keratin.

First introduced in the 1970s, many hair conditioners now contain silicone-based lubricants as their primary conditioning agents. Linear long-chain silicone copolymers applied to wet hair allow for quickened hair drying by their displacement of water molecules. They may also improve hair luster, help the appearance and texture of damaged hair, and reduce hair friction and static, thereby preventing a situation known as "fly-away hair." Enhanced ability to decrease hair static may be achieved by the addition of functional groups such as amines to the silicone molecules. Silicones commonly used include dimethicones, dimethicone copolyols, cyclomethicones, phenyl trimethicone, amodimethicones, and trimethylsilylamodimethicones.

Proteins and their hydrolysates are another class of conditioning agents; these include materials derived from animals and vegetables, such as keratin, collagen, elastin, silk, soy, corn, and wheat. They can be used to protect and repair damaged hair, including the temporary mending of "split ends."

Additional ingredients of conditioners include fatty compounds that provide hair lubrication and luster and act as thickeners, emulsion stabilizers and/or emollients (e.g., oleic acid, stearic acid, linoleic acid, lauryl alcohol, cetyl alcohol, stearyl alcohol, glycol stearate, polysorbate 20, beeswax, lanolin, glycerin, mineral oil, almond oil, and avocado oil). Solvents (e.g., water), pH adjusters (e.g., lactic acid, citric acid), botanical ingredients (e.g., rosemary, sage, ginseng), vitamins (e.g., B, A, E, panthenol, extract of wheat germ), UV absorbers (e.g., benzophenone-4), preservatives (e.g., parabens, disodium EDTA, diazolidinyl urea), color, and fragrance are also often added to provide product substance, ensure product chemical reactivity, enhance antioxidant properties, strengthen hair roots, stimulate natural growth, stimulate keratinization, protect against damaging sun-based UV radiation, ensure product antimicrobial purity, and enhance product quality.

HAIR GEL

While a true gel is a hydrocolloidal suspension, cosmetic products generally referred to as gels are semisolid in texture and usually clear to translucent in appearance. The viscosity of these products ranges from a thickened and readily pourable liquid gel to a semisolid stick. The product usually is in a thickened state within the squeeze tube or pump container and tends to decrease in viscosity during dispensation and application. Gel-type products may be water-based polymer gels or surfactant-based gels. Water-based gels usually contain alcohols to enhance clarity, product-thickening water-soluble polymers, resins, conditioning and moisturizing oils (e.g., lanolin, mineral oil, meadow-foam seed oil, sweet almond oil, safflower seed oil, castor oil), along with fragrance, preservatives (e.g., 1,3-bis[hydroxymethyl]-5,5-dimethylimidazolidine-2,4-dione [DMDM] hydantoin, tetrasodium EDTA, diazolidinyl urea, methylparaben, propylparaben, ethylparaben), antioxidants (e.g., panthenol, tocopherol), coloring agents, and UV stabilizers (e.g., benzophenone-4). Surfactant-based gels (also called microemulsion gels) have a high viscosity and usually contain surfactants (e.g., triethanolamine), oils, and product-clarifying polyols (e.g., propylene glycol, glycerol, polyethylene glycol). Gels may also contain fruit and/or herbal extracts (e.g., apple, apricot, grapefruit, orange, sage, euphrasia, ginger root, *Aloe barbadensis*, chamomile, jojoba, rosemary, tea tree leaf, lavender, peppermint) to enhance gel formation and product quality as a hair treatment.

The function of a modern hair gel is to bond the hairs together and maintain hair styling and/or to provide conditioning (usually with an oil or conditioning polymer). Bonding activity is attributable to the presence of film-forming adhesive materials such as polymeric resins or fixatives. The most commonly used resins for this purpose are the water-soluble nonionic polymers PVP and/or poly-*N*-vinyl formamide. Thickeners are often added to provide stiffness and holding power. One of the most widely used synthetic gelling/thickening agents is a polyacrylic acid resin cross-linked with the alkyl ether of materials such as sucrose (a carbohydrate) or propylene, commonly referred to as carbomers. Carbomers require chemical neutralization (e.g., using triethanolamine) if added with PVP to avoid the formation of an insoluble PVP-carbomer complex. Other thickeners used include hydroxyethylcellulose, hydroxypropylmethylcellulose, algae, hydroxypropyl guar (gum), acrylates, PEG compounds, hydrolyzed wheat starch, and polyquaternium-11. Acids such as glycolic, citric, lactic, and/or phosphoric acid are added as stabilizing agents, and agents such as oils, dimethicone, or glycerin are added to promote hair conditioning.

Because PVP has substantial humectant properties (and thus tends to absorb moisture in high humidity, causing the applied gel to become very sticky), alcohols (e.g., SD alcohol 40, ethanol, sorbitol) can be added to alleviate this problem. The addition of more humidity-resistant copolymers of PVP, such as vinyl acetate copolymer, allows the product to have increased resistance to moisture absorption. However, curl-activating gel products function by actually attracting moisture to the hair with the use of humectant polyols such as dimethylene copolyol.

Overall, hair gel should exhibit clarity in appearance, hair-holding capabilities under high humidity, pleasant odor, and the correct shear-thinning property to ensure even application of the product.

HAIR GROWTH PRODUCTS

Hair growth products first appeared in Japan in the 1950s, but not until 1979 was the first U.S. patent issued for a treatment for male baldness. In the 1980s, the market for these products developed extensively, and demand in the United States for these types of products continues to rise among both men and women.

Hair growth on the scalp and forehead of humans develops around the second month of fetal life. In the bottom basal layer of the epidermis, the primitive hair germ is formed. As the hair germ lengthens, the base of the hair follicle, also composed of epidermal cells, extends and surrounds mesenchymal cells of the dermis to form what is called the dermal papillae. Hair matrix cells (also epithelial cells) formed within the follicular base then proliferate and differentiate to grow into actual hair. Around the fifth month of fetal life, the follicles on the scalp continue to grow, while those follicles that supported hair growth on the forehead regress. The continual process of hair growth and loss involves a hair growth cycle that includes specific follicular stages of growth activity, intermediate regression, and resting (termed the anagen, catagen, and telogen phases, respectively). Hair that has ceased to grow during the telogen phase falls out naturally. Hair repeatedly grows and falls out in accordance with the hair cycle in a normal scalp, yielding approximately 90 percent of follicles in the anagen phase and 10 percent in the telogen phase at any given time. However, many biochemical and environmental factors may disrupt the normal hair cycle. If these factors shorten the anagen (growth) phase, they cause the hair to become fine and downy vellus hair resembling infantile hair called lanugo. This situation is typically manifested with the condition called male pattern baldness and can be exacerbated by the androgen hormone testosterone. When testosterone is converted into the more potent androgen dihydrotestosterone, it may then bind to hair follicle receptor target cells and initiate a cascade that accelerates hair loss. Should factors cause the telogen phase to lengthen, the result is the loss of a considerable amount of hair without immediate replacement. The scalp then becomes thin and noticeable balding results, as is the case with female pattern baldness (also called diffuse baldness). While decreasing mental stress and improving one's diet may assist in alleviating hair loss, many hair growth products often use active agents that increase cell viability, elicit antiandrogenic agents, inhibit the secretion of follicle-deteriorating sebum, and/or accelerate blood flow. A blood flow accelerant marketed in the United States and Europe as a hair growth agent is minoxidil. First introduced as a medicated treatment for high blood pressure (called an antihypertension drug), minoxidil acts by dilating blood vessels based on cellular potassium channel-opening action. As a secondary effect in many patients prescribed the drug, it was shown to produce a growth of hair anyplace on the skin containing hair follicles. While the specific mechanism of this hair growth promotion is unclear, minoxidil was approved by the Food and Drug Administration (FDA) in 1996 as an over-the-counter drug. Regular-strength hair regrowth treatment for men and separately marketed hair regrowth treatment for women each contains 2 percent minoxidil as an active ingredient. Extra-strength treatments for men contain up to 5 percent minoxidil. Inactive ingredients in these products include alcohol, propylene glycol (emulsifier), and purified water. While women's treatments are designed to regrow hair on the scalp, men's treatments are only designed to regrow hair on the top of the scalp, not for the treatment of frontal baldness or receding hairline. Used twice daily, product results may occur for some consumers within two to four months. Hair loss is usually manifested again with discontinued use.

HAIR REMOVAL PRODUCTS (DEPILATORIES)

A chemical depilatory is a preparation in the form of a liquid or cream that is used to remove unwanted hair from the surface of the body.

Depilatory literally means "to deprive of hair" (Latin *pilus*, hair). As far back as 4000 to 3000 BC, women were known to use a depilatory containing orpiment (natural arsenic trisulfide), quicklime (used to make cement), and starch made into a paste. Around 500 BC, women made depilatory creams from medicinal drugs, such as bryonia. In the middle of the first century AD, Romans used depilatories with ingredients such as resin, pitch, white vine or ivy gum extract, ass's fat, she-goat's gall, bat's blood, and powdered viper. During Elizabethan times, walnut oil, quicklime, or mixtures of vinegar and cat's dung were applied to remove hair. In the early eighteenth century, American women applied poultices of caustic lye to burn hair away, and by the middle of the nineteenth century, powdered depilatories were marketed throughout the United States. In 1940, a New York-based company developed America's most successful depilatory lotion as a result of wartime shortages of razor blade materials. The product included the active ingredient calcium thioglycolate, which destroys protein hair structure and reduces hair texture and strength.

Modern chemical depilatories are highly alkaline pastes, gels, aerosols, roll-on forms, creams, or lotions. The pH of these depilatories is usually between 10 and 12.5. They contain alkali or alkali-earth sulfides (usually up to 35 percent sodium, barium, or strontium sulfide) or mixtures of alkali-earth hydroxides (usually 5 to 10 percent potassium or calcium hydroxide), along with salts of aliphatic mercapto acids (usually 2 to 5 percent sodium or calcium thioglycolate, or thioglycolic acid). Combined, these overall alkaline active ingredients destroy some of the peptide bonds within the fibers of keratin protein that constitute each shaft of hair. When used in a product, the concentration of calcium thioglycolate $[Ca(CH_2SHCOO)_2]$ is generally maintained to yield desired results within a reasonable time frame (five to fifteen minutes) depending on hair coarseness, while avoiding the potential to injure the skin. The alkaline pH can possibly damage and degrade dermal proteins, causing skin irritation, excess exfoliation, or allergic contact dermatitis. Fortunately, this dermal effect is usually temporary, lasting only hours or a few days. It is also recommended that consumers read product labels and select the formulation appropriate for the intended use, as skin sensitivity varies throughout potential application body areas (e.g., legs, bikini line, underarms, face). Acting like a chemical razor blade, chemical depilatories cause degradation of the hair keratin and deterioration of hair fibers to a gelatin-like mass. Because the protein structure of the hair is destroyed, the hair will easily separate from the skin surface and be manually removed by wiping or scraping. This hair removal process works best on soft, fine hair rather than coarse hair, and although the results vary with the individual, a few days to two weeks is the usual duration of the hair-free period.

Other chemical ingredients often included in commercial chemical depilatories are water, moisturizers and emollients (e.g., mineral oil, almond oil, sunflower seed oil, jojoba oil, lanolin, glycerin, propylene glycol, polysorbate 20), emulsifiers (e.g., cetearyl alcohol, ceteth-20, ceteareth-20, PPG-15 stearyl ether, sodium lauryl sulfate), antioxidant vitamins (e.g., tocopheryl acetate [vitamin E acetate], ascorbic acid [vitamin C]), binders and thickeners (e.g., silica, magnesium silicate, xanthan gum, copolymer, acrylates), fragrance, skin-soothing herbal extracts (e.g., *Aloe vera*, *Camellia oleifera*, *Anthemis nobilis* flower, panax ginseng root), color pigments (e.g., titanium dioxide, iron oxide, chromium hydroxide green), antiseptic oils (e.g., bisabolol), chelating agents (e.g., sodium gluconate), and preservatives (e.g., urea).

HAIRSPRAY

The concept of a preparation to control the appearance of hair is quite an historical notion. Originally, the preparation was probably animal fat (or an emulsified version that led to the creation of hair cream), followed by the use of various gums and shellac lacquer by female populations. While the concept of an aerosol originated in eighteenth century France, scientists during World War II developed a small aerosol can pressurized by a liquefied gas (e.g., fluorocarbon). It was this design that made products such as hair spray possible, along with the work of an inventor in the 1950s that involved the creation of a crimp-on valve for dispensing gases under pressure, with the additional creation of the first clog-free valves for spray cans.

The ideal fixative allows the hair to flow naturally while remaining in style, resists high humidity while not being brittle, and is easy to comb. The fixative should also confer a good high gloss (e.g., in the "glassy" regime), not flake onto shoulders like dandruff when dry combed after application, and be readily removed upon washing.

Modern hairspray consists of a solution of long, chainlike molecules (called polymers) in a highly volatile solvent. Some brands may also contain oils such as resins and lanolin. In general, a volatile substance is one whose state is unstable at room temperature and may readily change from liquid to gas form. Thus, hairspray is in liquid form within the can, as air pressure has been removed. The can is frequently composed of compounds (e.g., aluminum monobloc or tri-layered steel) that allow for a decreased likeliness of puncturing. Spraying the product results in the deposit of a polymer layer around each hair after evaporation of the volatile solvent. The web of polymer molecules on the hairs yields a stiff texture and allows the hairs to resist changing shape.

A solvent once popularly used was a compound of carbon, fluorine, and chlorine (a chlorofluorocarbon [CFC]). In general, CFCs are considered nearly optimal aerosol propellants, having both nontoxic and nonflammable chemical characteristics. However, when it was discovered by scientists in the late twentieth century that their ubiquitous use and release into the atmosphere led to the destruction of the stratospheric ozone layer, they were eventually replaced with other solvents (e.g., alcohols and hydrocarbons). Unfortunately, hydrocarbon propellants in combination with alcohol are extremely flammable, and consumers are warned to avoid heat, fire, and smoking during use until sprayed hair is fully dry. Thus, aerosol preparations are currently manufactured to decrease volatile organic compound content while also decreasing flammability performance.

In general, the characteristics of polymers change with temperature and chemical environment. PVP was one of the first hairspray polymer resins used, and it is also used to glue the layers of wood in plywood together. Because PVP is water soluble, absorbing atmospheric moisture and becoming sticky, water-insoluble polymers (e.g., vinyl acetate or polydimethylsiloxane) may be added, which allows the spray to dry to a brittle film for a longer-lasting hold. Other types of polymers used in plastic-based hairsprays are copolymers with vinyl acetate and/or maleic anhydride. In addition, some "natural" hairsprays use water-soluble hair-stiffening herbal polymers such as gum arabic (found in the sap of *Acacia senegal* trees of Sudan), gum tragacanth (*Astragalus gummifer*), a gum also used to stiffen calico and crepe, or karaya gum (*Sterculia urens*).

Hair-holding and -stiffening resins are also available as mousses (foam or froth). As in hairsprays, the active ingredient is a resin such as PVP. In addition, silicone polymers may be added to provide a sheen or shiny texture to the hair.

LIPSTICK

Lipsticks first appeared in the ancient city of Ur, near Babylon, 5,000 years age. Egyptians made lipsticks from finely crushed carmine beetles, which yielded a deep red pigment, mixed with ant eggs as a base. Elizabethan lipsticks were a blend of cochineal and beeswax. For many centuries, lip reddeners consisted of a very poisonous substance called mercuric sulfide. Lipsticks in their modern form were introduced after World War I; they were colored with a dye and acid-base indicator (carmine) chemically extracted from the small red cochineal insect (*Coccus cacti*). Indelible long-lasting lipsticks were introduced in the 1920s and became colored upon reacting with the skin. It is this technology of the 1920s on which modern color-change lipstick chemistry is based.

Modern lipstick is formulated to provide both protection for the delicate tissues of the labia and color for appearance. The chemical composition of lipstick varies greatly among manufacturers. However, lipstick in general is formulated to provide uniform color coverage while being neutral in taste, stable under normal fluctuations of temperature, moisture, and air flow, and lacking toxicity and irritancy. Lipstick should remain relatively stiff and intact within the dispenser tube but flow easily onto the lips under application pressure. To achieve these goals, the specific chemical composition may include a mixture of oils, waxes, pigments, emollients, antioxidants, and preservatives. Perfumes are often added to mask the unpleasant fatty odor of the oil. Much of the body mass of the lipstick is composed of a mixture of a nonvolatile oil (e.g., castor, vegetable, mineral, or lanolin) and solid wax (e.g., beeswax or carnauba). The addition of oil allows the wax-based product to be softened and easily applied. In addition, esters of fatty acids (e.g., 2-propyl myristate) are commonly added to reduce any "stickiness."

Many consumers view the color of lipstick as the most important characteristic of this product. The colors and dyes of lipsticks are generally regulated within the United States and include many water-insoluble (oil-soluble) products, such as brilliant blue, erythrosine, amaranth, rhodamine, tartrazine, dibromofluorescein, and tetrabromofluorescein (bluish-red compound). The dyes must be water insoluble; otherwise, the color would quickly fade or be removed in a short time by the consumer through the movement of the saliva-soaked tongue across the lips. Water-soluble dyes such as green or blue food dyes may be used to provide lipstick coloration, but they are usually first "laked" or combined with metal oxides such as aluminum hydroxide $[Al(OH_3)]$ to form an insoluble precipitate that is then suspended in the oil base of the lipstick.

Eosin is a commonly used pigment that yields an intense red color upon application after chemical reaction with numerous $-NH_2$ (amine)-containing skin surface proteins. For some color-change lipsticks, the lightly colored eosin pigment is masked in the applicator tube by a laked green or blue dye. Thus, the laked dye is the color viewed before the lipstick is applied, and the color change when the lipstick is applied to the lips is intense as eosin yields a red color. These types of lipsticks frequently contain acids (e.g., citric or lactic), as the color-change dyes tend not to function properly under basic conditions.

The very popular "mood lipsticks" are composed of weak acid color pigments that have a conjugate base form with a remarkably different color. Thus, these lipsticks are simple acid/base indicators. When the lipstick reacts with the skin, it changes to a color varying according to the pH of the skin at the time of application. Since the pH of the skin is dependent on numerous physiological factors (e.g., physical activity level, stress, nutrition, hormonal fluctuations) and genetically different natural base skin colors, the color of the mood lipstick varies from application to application.

LIP TREATMENTS

Modern lip treatments are formulated to moisturize and protect lips exposed to potential drying and chapping from environmental factors such as solar UV radiation, wind, and cold temperatures. Important base materials used in the development of lip treatments include waxes (e.g., candelilla [*Euphorbia cerifera*], carnauba [*Copernicia cerifera*], Japan, ceresin, microcrystalline, ozokerite, beeswax [*Cera alba*], paraffin), oils (e.g., castor seed [*Ricinus communis*], corn, safflower, coconut, hydrogenated castor, sunflower seed [*Helianthus annuus*]), silicone derivatives (e.g., dimethicone, dimethicone copolyol, C30-40 alkylmethicone), and lanolin derivatives. In combination, these ingredients provide high shine, rigidity, hardness, enhanced wear, pleasant application feel, and moisturizing effects. In addition, other chemicals, such as fatty acid esters (e.g., isopropyl myristate, palmitate), hard butters (e.g., shea butter [*Butyrospermum parkii*], cocoa butter [*Theobroma cacao*]), vitamins (e.g., magnesium ascorbyl phosphate [vitamin C], tocopheryl acetate [vitamin E], panthenol [provitamin B5], retinyl palmitate [vitamin A], tocopherol [vitamin E]), herbal extracts (e.g., *Aloe vera*), emulsifiers (e.g., petroleum, squalane, allantoin, cetyl alcohol), and skin-soothing organic compounds (e.g., menthol, phenol, camphor), provide additional product texturization and skin protection. While flavoring extracts (e.g., natural orange, vanilla fruit [*Vanilla planifolia*]) and fragrance provide pleasant product odor and taste, the addition of synthetic preservatives (e.g., methylparaben, propylparaben, BHT) assists in maintaining product purity and freshness with repeated use.

Many consumers desire the additional lip treatment feature of protection from damaging type A and/or B solar UV radiation. Added active sunscreen ingredients must be safe when ingested in small amounts and not interfere with product esthetic characteristics. These chemicals usually include UV-B protectors such as octyl methoxycinnamate and padimate O and UV-A protectors such as oxybenzone added at concentrations ranging from 7.5 to 1.5 percent.

LIQUID FOUNDATION MAKEUP

The first liquid foundations were known as neck and arm whiteners. They were developed for use in the theater and were early attempts to improve the characteristics of applied face powder. Early pharmaceutical methods used to develop improved foundations included combinations of calamine lotion, zinc oxide, glycerin, and water. Until the 1940s, foundations were in the form of nonflowing and very greasy pastes and creams (often called "grease paint"). The cosmetic industry soon thereafter used different types of emulsifiers and thickeners combined with wetting agents, vegetable oils, and plasticizers to improve the formulations of spreadable viscous lotions.

The modern makeup base, in liquid, cream, or cake form, must provide a quality finish, adequate coverage, and appropriate pigmentation. Consumer selection of foundation makeup should be based on skin type (e.g., presence of scars, enlarged pores, uneven skin tones, etc.), degree of sebum production ("oiliness"), skin color (e.g., sallow yellow, ruddy red, or neutral), and depth of skin color (light, medium, or dark). General types of materials used in most foundation makeups may be classified as part of the powder phase, aqueous phase, polyol phase, or oil phase.

The powder phase includes a prominently used skin covering and opaque materials such as titanium dioxide and/or zinc oxide, followed by fillers including kaolin, talc, mica, and sericite (the hydrated form of mica), which create a matte finish on the skin. In addition, metallic stearates (e.g., zinc stearate, aluminum stearate) are commonly added to provide both opacity and water repellency. Stable and colorfast inorganic colorants used include red, yellow, and black iron oxides, ultramarines, along with certain organic lakes (e.g., D&C Red, D&C Orange). To obtain a pearlescent appearance, materials such as bismuth oxychloride and mica are added. The aqueous phase includes mostly water, followed by wetting or dispersing agents that assist in pigment blending (e.g., lecithin and its derivatives, sulfosuccinates, ethoxylated sorbitan esters). Also found in the aqueous phase are gums and clays that operate as co-emulsifiers, thickeners, and pigment-suspending agents (e.g., magnesium aluminum silicate, bentonite, sodium carboxymethylcellulose, methylcellulose, hydroxyethylcellulose, xanthan gum). The polyol phase usually consists of materials that improve makeup spreadability, decrease drying rate, aid in freeze-thaw product stability, and increase product smoothness and creaminess (e.g., glycol, butylene glycol, glycerin, polyethylene glycol). The oil phase is composed of both liquid and solid materials that assist with product emulsification, spreadability, pigment dispersion, and viscosity control. These oil phase products include fatty acids (e.g., stearic acid, isostearic acid, oleic acid), waxes (e.g., carnauba, ozokerite, candelilla, microcrystalline, beeswax, ceresin), liquid hydrocarbons (e.g., isohexadecane, isoeicosane, hydrogenated polybutanes), mineral oil, vegetable oils (e.g., sesame, peach kernel, apricot kernel, avocado, macadamia, kukui, meadow-foam), and pastelike butters (e.g., shea, karate, cocoa). Emulsifying and thickening agents (e.g., cetyl esters, propylene glycol, glyceryl esters, sorbitan, ethoxylated sorbitan esters and their derivatives), along with antioxidants (e.g., tocopherols, BHT), preservatives (e.g., methylparaben, propylparaben, ethylparaben, butylparaben, urea, trisodium EDTA), and fragrance added within the oil phase, may improve application properties and product quality and purity.

Oil-free makeup brands possess few emollient ingredients and tend to use fatty acid esters (e.g., isostearyl palmitate, isostearyl stearoyl stearate, cetearyl octanoate) as replacement for oil phase components. In addition, oil-absorbing ingredients such as porous silica, nylon-12, talc, and kaolin are added to control daily oil production. Extended-wear makeup achieves the product goal with the use of a rosin ester (e.g., pentaerythrityl hydrogenated rosinate). Antiwrinkle makeup may contain ingredients such as C2-C28 α-hydroxy carboxylic acid or rosemarinic acid that have been shown to be effective in treating fine lines and wrinkles, or spherical material to diminish the appearance of such skin imperfections.

For specialized long-lasting water-in-silicone foundation formulations, typical oil phase materials are replaced by organofunctional silicones (e.g., cyclomethicone, stearyl dimethicone, cetyl dimethicone), silicone emulsifiers (e.g., dimethicone copolyol), suspending agents (e.g., quaternium-18 hectorite), and thickening agents (e.g., hydrogenated castor oil, microcrystalline wax). The powder phase usually incorporates pigment and filler materials treated with silicone materials (e.g., dimethicone, methicone) and metallic soaps (e.g., myristyl myristate, aluminum myristate) to aid in pigment and filler dispersion. Water, preservatives, and electrolytes are major components of the aqueous phase.

MASCARA

Egyptian women from the First Dynasty developed the art of decorating the eyes by blackening the lashes and upper lid with kohl, a preparation made from antimony or soot. By the middle of the first century AD, this technique was also widely used by the Romans. By the beginning of the twentieth century, mascara was commercially developed by cosmetic businesses and universally marketed.

Modern mascara has a base of soap (the majority of product), oils (e.g., lanolin), waxes, and fats. Mascara is colored and darkens the eyelashes by the presence of pigments such as iron oxide (brown), carbon or lampblack (black), chromium(III) oxide (Cr_2O_3) (green), titanium dioxide (TiO_2), or a silicate that contains sulfide ions named ultramarine (blue). Despite the term "lengthening mascara," these types of marketed mascaras simply darken the ends of the lashes for increased contrast, allowing the lashes to have a longer appearance. While a typical composition is approximately 40 percent wax, 50 percent soap, 5 percent lanolin, and 5 percent coloring pigment, mascara may also contain shellac, quaternary ammonium, alcohols, antimony, nickel, antibacterial agents, or colophony (rosin) derived from coniferous trees, which yields good tackifying properties and increased water resistance.

Allergic reaction to mascara ingredients (e.g., colophony) is of concern to product manufacturers and regulators. More serious concerns, however, are eye infections caused by bacterial contamination from consumer-shared eye makeup and/or regular use of eye makeup more than three months of age as well as potential corneal damage from misuse of the mascara wand applicator.

MOUTHWASH

A mouthwash may be defined as a nonsterile aqueous solution used mostly for its deodorant, refreshing, or antiseptic effect. Mouthwashes or rinses, when used as an adjunct to regular oral hygiene methods such as flossing and brushing, are designed to reduce oral bacteria, remove food particles, temporarily diminish acute halitosis, and provide a pleasant oral aftertaste. While the above functions are classified by the FDA as those of cosmetic rinses sold over the counter, FDA-classified therapeutic rinses sold by prescription or over the counter contain added ingredients that assist in protecting the consumer against some oral diseases and are approved by the American Dental Association.

Mouthwash ingredients vary, but common active ingredients in antiseptic rinses include alcohol, chlorohexidine gluconate, cetylpyridinium chloride, benzoic acid (acts as a buffer), or hexidine. In addition, many contain essential oils that possess antibacterial properties, including phenol, thymol, or eugenol. Some rinses contain high levels of alcohol, ranging from 18 to 26 percent. This may lead some consumers to sense burning in the cheeks, teeth, and/or gums. In addition, intoxication may result if alcohol-containing rinses are swallowed or used excessively and is potentially lethal in children exposed to small doses. Chlorohexidine gluconate is quite effective as an antimicrobial agent and an inhibitor of plaque and calculus formation, but it may stain teeth in extended use. Cetylpyridinium chloride has bactericidal in addition to antiplaque activities. In general, oral antiseptic agents aid in reducing bacterial growth that leads to plaque formation but do little to prevent oral diseases such as gingivitis and periodontal disease. Rinses of any kind are not recommended as substitutes for regular dental professional examinations and proper home care, including brushing with fluoride toothpaste and flossing. This recommendation is warranted because chronic halitosis and an unpleasant oral taste are potential indicators of oral diseases.

Some other active ingredients in mouthwashes that assist in fighting plaque and halitosis are triclosan (an antiplaque agent with anti-inflammatory properties), peroxide and perborates (antiplaque and oxygenating agents), and chlorine dioxide (reduces halitosis by reducing the concentration of volatile sulfur compounds). Mouthwashes may also contain chemicals such as glycerin (provides substance to formula and acts as a humectant), sweeteners (e.g., sodium saccharin), buffers (e.g., sodium benzoate), flavors, coloring, and emulsifiers (e.g., polysorbate 80) that help stabilize flavor oils in mouthwash base. Interestingly, rinsing with a fluoride mouthwash has been shown to provide benefit to oral health over rinsing with water alone or with a nonfluoride mouthwash. As fluoride has been shown to strengthen tooth enamel, fluoride mouthwash may be used by consumers with an excessive tendency toward tooth decay or by those living in areas with inadequate fluoride in the water supply.

Nutritional supplements have recently been included in some brands of non-alcohol-based mouthwashes marketed as "natural products" and may include supplements absorbed through the gums during consumer use, such as folic acid (reduces gum inflammation and bleeding and binds to toxins secreted by plaque-forming bacteria) and zinc (antiplaque and antibacterial properties). Myrrh and clove oils are often added to these products, as they may have antibacterial and anti-inflammatory properties along with controlling halitosis.

NAIL POLISH

Nail polish can be historically traced back approximately 5,000 years, to at least 3000 BC, when it originated in China. During the Ming Dynasty, Chinese nail varnishes and lacquers were synthesized from a combination of beeswax, egg whites, gelatin, vegetable dyes, and gum arabic. The Egyptians were known to use orange henna to stain their fingernails. In China as well as in Egypt, color symbolized particular social classes. During the Chou Dynasty (600 BC), gold and silver were the royal color choices, and later, royalty preferred black and red and thus applied these colors to their nails to indicate their status.

Modern nail polish is actually a refined version and variation of automobile paint. The polish, designed to protect and beautify nails, is a highly specialized and flexible lacquer that does not easily crack and flake with natural nail movement. The formulation goal of most manufacturers is to provide as much film as possible in a single application coat while retaining application ease, quick drying time, maximum hardening, chip resistance, and a natural pearl essence. Usually, the application of two coats is necessary to obtain adequate film thickness and sufficient opacity. The three major ingredients in most nail polish brands are organic solvents, resins (thickeners or hardening agents), and color pigments. The most common organic solvents are ethyl acetate and butyl acetate (both also used as solvents in nail polish removers). As volatile solvents, these esters (synthesized by reacting a carboxylic acid with an alcohol; the general formula is R-COO-R') evaporate quickly, leaving the resin/pigment mixture attached to the nail surface as a thin coating. Other commonly used solvents include acetone, toluene, methyl chloroform, dipropylene, ethyl alcohol, and isopropyl alcohol. Solvents are responsible for the strong odor of nail polishes.

Resins, types of polymers, are the thickening and hardening agents that, without pigments, serve as colorless nail protectors resembling clear furniture lacquer. These agents include nitrocellulose (collodion) and different acrylate and polyester/polyurethane copolymers. Copolymers include chemicals such as methacrylic acid, isobutyl methacrylate, toluenesulfonamide formaldehyde resin, phthallic anhydride/trimellitic anhydride/glycol copolymer, tosylamide/formaldehyde resin, and dimethicone copolyol.

Nail polish pigmentation (coloring) tends to be the essence of the polish and of paramount importance to the consumer. A variety of D&C laked dyes (drug and cosmetic dyes approved by the FDA) are used in combination to achieve the desired color. Coloring may also be attributable to the presence of chemicals such as chromium oxide greens, chromium hydroxide green, ferric ferrocyanide, ferric ammonium ferrocyanide, stannic oxide, titanium dioxide, iron oxide, carmine, ultramarines, and manganese violet. Sparkling and reflective particles such as mica, bismuth oxychloride, natural pearls (guanine), and aluminum powder are used to make "frost" and "shimmer" polishes appear glittery or pearl-like.

Other ingredients in nail polish include plasticizers (e.g., dibutyl phthalate, camphor, citrates, adipates, glycol dibenzoate) that serve as molecular lubricants, allowing the resin to remain flexible after drying, and increase resin resistance to oil and water. Dispersants (e.g., organically modified clays such as stearalkonium bentonite and stearalkonium hectorite) are additives that control flow by helping the pigments mix with the resin and solvent, thereby preventing sinking of the color particles. Ultraviolet stabilizers (e.g., benzophenone-1) may be added to prevent the polish from changing color after excessive UV sunlight exposure. In addition, chemicals such as colorant consistency regulators (e.g., palmitic acid) and antioxidant preservatives (e.g., citric acid) may be added.

NAIL POLISH REMOVER

Most commercial nail polish removers consist mainly of volatile organic solvents such as ethyl acetate ($CH_3COOC_2H_5$), a common solvent for the polish itself, or acetone (CH_3COCH_3). Acetone is the simplest of ketones (a class of organic compounds) and is also known as dimethyl ketone, propanone, or β-ketopropane. It is a colorless liquid with a distinct smell and taste. It evaporates easily, is flammable, and dissolves completely in water and organic solvents. Since it may be used to dissolve other substances, care must be used when applying acetone-based nail polish remover to the nails because the product may also remove nearly any other type of paint or varnish. Acetone may also dissolve the surface layer of linoleum and most floor tile. As well as its use as a nail polish remover, acetone has many industrial uses. For example, it is used as a gelatinizing agent in explosives and in the manufacture of plastics, fibers (e.g., rayon), and drugs and other chemicals (e.g., rubber cement and some cleaning fluids). While acetone is a manufactured product, it is also found naturally in plants, trees, volcanic gases, forest fires, and as a product of the breakdown of body fat. It is also present in vehicle exhaust, tobacco smoke, landfills, and hazardous waste sites. A large percentage of the acetone released during its manufacture or use enters the atmosphere, and approximately half of the total amount is degraded via sunlight or other chemicals during the period of a calendar month. It can move from the atmosphere into water or soil via precipitation, where microorganisms may eventually break it down. Human exposure to a small amount of acetone will result in its chemical breakdown by the liver. However, breathing moderate to high levels of acetone, a known neurotoxin, for short periods of time may cause nose, throat, lung, and/ or eye irritation, headaches, confusion, increased pulse rate, nausea, vomiting, and possibly coma. Swallowing very large amounts of acetone may result in unconsciousness and damage to skin within the mouth.

Because acetone contact with the skin may result in irritation and dermal damage, nail polish removers containing acetone usually also contain emollients such as mineral oil, castor oil, or lanolin to prevent the nails and surrounding skin from becoming dry and devoid of natural oils. These oils also decrease the evaporation time of the volatile paint-removing solvents, thereby allowing more time for the product to work efficiently.

Nail polish removers may also contain ingredients such as general solvents (e.g., propylene carbonate), products found in commercial paint strippers (e.g., dimethyl glutarate, dimethyl succinate, and dimethyl adipate), emollients and moisturizers (e.g., glycerin and panthenol), preservatives (e.g., propylene glycol), vitamins (e.g., dl-α-tocopheryl acetate [vitamin E]), fragrance, and coloring dyes. Some brands also contain a chemical called denatonium benzoate ($C_{28}H_{34}N_2O_3$), one of the most bitter-tasting substances known. Since most humans have a natural aversion to ingesting highly bitter chemicals, this substance is often added to toxic household liquid products to decrease accidental poisoning via swallowing of substantial amounts.

PERMANENT WAVING OF HAIR

The desire of individuals to improve the natural appearance of their hair can be dated back to ancient Egyptian and Greco-Roman societies. Such dissatisfaction continues to be the driving factor in the development of modern permanent waving by the cosmetics industry. The first permanent waving systems were pioneered in the early 1900s and included a nearly nine-hour procedure using permanent waving machines and various solutions. In the 1920s, permanent waving as a chemical method was invented. In the 1930s, scientists demonstrated that disulfide bonds that give proteins their spatial three-dimensional structure could be severed at ambient temperature and slightly alkaline pH with the action of sulfides or mercaptans, and by the 1940s, the use of thioglycolic acid as a "cold wave" treatment revolutionized the "perm" industry and provided a foundation for the chemistry of modern permanent waving methods.

Hair is composed primarily of the protein keratin. As a protein macromolecule, keratin exists as multiple amino acid chains linked together by forces that include hydrogen bonds, salt bridges, disulfide linkages, and hydrophobic interactions. While hydrogen bonds between protein chains may be disrupted by water and salt bridges destroyed by alterations in pH, disulfide linkages may be broken and restored when hair is permanently waved or straightened. In keratin, the amino acid cystine dominates the polymer, and the sulfur-sulfur bonds of cystine residues provide hair strength by connecting parallel strands of protein.

Chemical modification may be achieved by wetting the hair alone, thereby protonating and breaking the hydrogen bonds that hold the hair proteins in place. Drying the hair while held in a straight or curlier shape than before wetting sets the new hairstyle temporarily by the presence of newly realigned polymeric chains. A permanent wave change is achieved by applying chemicals to the hair that break not only the hydrogen bonds but also the disulfide bonds. These disulfide bonds are then reconstructed when the hair is in curlers, allowing the hair to permanently take on the curled shape.

When the hair is placed in curling rollers, permanent wave lotions containing a mercaptan reducing agent (e.g., thioglycolic acid [HS-CH_2COOH]) that breaks the disulfide bonds are applied. This agent effectively transfers the hydrogen from its own sulfur to the cystine units and cleaves them into cysteines. Thioglycolic acid is usually applied in a suitable pH buffer (usually slightly basic), together with a hair cuticle softener. This process is usually accompanied by a slight smell of rotten eggs. The hair is then treated with a mild oxidizing agent (e.g., a dilute solution of hydrogen peroxide [H_2O_2]) that neutralizes the effect of thioglycolic acid and allows the covalent disulfide linkages to be reformed between free cysteine bases to form cystine-cystine bonds. The newly reorganized strands are now in their new set "curled" positions to provide hair with a different shape than before the chemical treatment. These reorganized disulfide bonds remain intact whether the hair is subsequently wet or dry. However, since the oxidation process does not result in the reformation of all of the disulfide linkages cleaved during the reduction step, permanent chemical treatment may result in hair weakening and loss of tensile properties. Permanent wave curls naturally grow out as new hair is formed. Other types of "thio-free" permanent wave solutions contain active reducing agents such as ethanolamine sulfite, cysteamine HCl, ammonium thiolactate, or monoethanolamine thiolactate. An alkaline compound such as ammonium hydroxide usually neutralizes such thio-free wave solutions. Other types of neutralizing agents used in permanent wave formulations include perborates, ammonium persulfate, and sodium or potassium bromate.

SHAMPOO

The average human scalp contains approximately 125,000 to 150,000 hair fibers. Hair is nonliving material composed mostly of the fibrous protein keratin. Sequential bundling of keratin fibrils leads to the formation of the main body of the hair fiber, the cortex. Surrounding the cortex are lifeless cells known as cuticle cells that allow for protection by forming a sheath around the hair. At the center of this structure is the hair shaft. Coiled proteins surrounding the shaft provide hair with the properties of elasticity and bounce and contain melanin pigment that provides hair coloration. While the hair shaft extends outward from pores of the upper epidermis layer of the skin, the deeper dermis skin layer contains the hair bulb and root located within the hair follicles, the structures responsible for hair growth. Located adjacent to each hair follicle is a sebaceous (oil) gland, which produces and releases sebum. Sebum coats the cuticle cells of the hair and also aids in maintaining hair flexibility, shine, and gloss by keeping the cuticle sheath lying flat and preventing dehydration.

As a protein, keratin consists of amino acid chains sustained by different types of forces, including hydrogen bonds, salt bridges, disulfide linkages, and hydrophobic interactions. Of these forces, hydrogen bonds and salt bridges are the most important when considering the action of shampoo, as hydrogen bonds between protein chains are disrupted by water and salt bridges are destroyed by changes in pH. A strand of hair is most stable and strongest at a pH of 4 to 6 (slightly acidic), as the maximum number of salt bridges exists at a pH of approximately 4.1. Under basic conditions, the cuticle cells tend to become unstacked, and this dislodging of the protective tile cells leads to reduced light reflection and the appearance of hair lacking luster. In addition, the raised cuticle cells may also lead to tangling and the escape of cortex moisture leads to dry and brittle hair. Thus, the control of pH (between pH 4 and 9) is paramount when considering the chemistry of hair care products.

Rinsing hair with water alone will cause the keratin to become more stretchable by absorbing water and softening. However, adding a synthetic cleansing agent such as those found in shampoo to the hair during water rinsing will also allow for the removal of the water-insoluble oily sebum and associated debris from the hair shaft. Detergent action associated with shampoos is the result of surfactants. Surfactant is an abbreviation for "surface-active agent." Surfactants possess hydrophilic (water-soluble) heads and hydrophobic (water-insoluble) fatty acid tails. The water-soluble head of the surfactant may carry a negative charge (anionic surfactant), a positive charge (cationic surfactant), may lack any charge (nonionic surfactant), or the surfactant molecule may carry both a positive and a negative charge (amphoteric surfactant). Most major surfactants used in modern shampoos include all but cationic surfactants because of their skin-irritating potential. Surfactants promote the solubilization of sebum-based oil and dirt in water by lying across the oil-water interface and emulsifying the oil droplets, thereby allowing the oily soil to be removed from the hair via water rinsing.

The word "shampoo" is thought to come from the Hindu word *champo*, which means "to massage" or "to knead." The first successful retail shampoo was developed in 1930. Until that time, and through World War II, the cleansing agent in shampoos was soluble soap (usually sodium and/or potassium salts were preferred). Soap-based shampoos readily formed insoluble calcium and magnesium salts in the presence of hard water, thereby leaving a dulling film on hair. Consumers would often apply a rinse containing vinegar or lemon juice to remove the "scum" film.

The most common primary surfactants used in modern shampoos are the lauryl sulfates and lauryl ether sulfates. The preferred counter-ions for these anionic surfactants are sodium, ammonium, or potassium to confer good water solubility. Cosurfactants, which are frequently included for their ability to enhance foaming power, include chemicals such as coco monoethanolamide, lauric acid diethanolamine, coco diethanolamide, cocamidopropyl betaine, and other betaines (e.g., lauramidopropyl betaine). The foaming ability of shampoos may be additionally boosted by adding compounds such as fatty acid alkanolamides. Shampoos designed for oily, normal, and dry hair seem to differ primarily in the concentration of surfactants (e.g., oily hair: more concentrated detergent; dry hair: more dilute detergent). Amphoteric surfactants are often used to allow for skin mildness (e.g., in shampoos designed for infants) and include chemicals such as sodium alkyl amphoacetates, sodium 3-dodecylaminopropionate, sodium 3-dodecylaminopropane sulfonate, and N-alkyltaurines.

Other functional ingredients included in modern shampoos are those that confer benefits other than cleaning, such as agents to provide moisturizing body (e.g., proteins such as keratin, collagen, silk protein, hydrolyzed soy protein, hydrolyzed wheat protein), conditioning (e.g., oleyl alcohol, glycerine, stearyl stearate, glycerides derived from natural plant and animal oils, shea butter, allantoin, *Aloe vera* gel, dimethicone, and cationic polymers such as polyquaternium-10 and polyquaternium-7), antistatic properties (e.g., cationic detergents such as trimethyl alkyl ammonium chloride, tricetyl methyl ammonium chloride), UV protection (e.g., benzophenone-3), or antidandruff properties (e.g., piroctone olamine, zinc pyrithione). In addition, modifiers are added to confer desirable flow properties and to stabilize suspensions of insoluble particulate components (e.g., volatile silicones, carbomers [cross-linked polyacrylic acids], acrylates, cellulosics [e.g., hydroxypropylmethylcellulose], xanthan gum, glycol stearate, cetyl palmitate, glyceryl distearate, sodium sulfate).

Solvent fillers (e.g., water, propylene glycol), preservatives (e.g., formaldehyde, DMDM hydantoin, BHT, methylisothiazolinone, methylchloroisothiazolinone, parabens), pH adjusters that enhance foaming action (e.g., triethanolamine, myristic acid, sodium chloride, citric acid, lactic acid), chemical sequestrants/ligands (e.g., EDTA), herbal extracts (e.g., lavender, bergamot, rosemary, peppermint, soapwort, yucca, Iceland moss, orange flower, lemongrass, grapefruit seed, cucumber, watercress, parsley, matricaria, mushroom, chamomile, jojoba, fennel seed, golden seal), vitamins (e.g., panthenol, ascorbic acid, retinyl palmitate, riboflavin, tocopheryl acetate), colors (e.g., caramel, henna, D&C colors, FD&C colors), and fragrance also can be added to enhance the composition, purity, chemical effectiveness, and aesthetic quality of the product.

SHAVING CREAM PREPARATIONS

Throughout recorded history, the act of removing hair from various parts of the body via shaving with razors has been associated with such factors as cleanliness and hygiene, social rank, military dominance, fashion trendiness, and vanity. Early shaving preparations often included soaps, oils, or herbal treatments such as cherry laurel water. Most modern shaving preparations are designed to lubricate the skin and allow the cutting blade to cut the protruding hair shaft (and not the surrounding skin), to moisten and soften hair to ease cutting capabilities, and to serve as a demarcation of where shaving has already occurred, thereby preventing skin irritation with repeated razor strokes.

While the use of water alone can soften hair, the addition of a fatty acid salt (i.e., soap) improves the performance of water as a shaving aid by creating a film on the skin. This film created by shaving soaps allows the blade to slide with decreased resistance along the outer epidermal skin surface without cutting into the vascular (blood vessel-possessing) deeper dermal skin layer. In general, soap results from the treatment of natural oils (e.g., animal tallow, fish oil, or sunflower oil, vegetable oils such as corn oil, soybean oil, olive oil, palm oil, coconut oil) and fats with an alkaline solution followed by neutralization. This process allows for the production of free fatty acids and glycerine via hydrolysis of the fatty acid esters. The texture of the soap is dependent on the fatty acid chain length and the degree of saturation. For example, whereas "hard" soaps result from the use of saturated fatty acids (e.g., stearic acid, palmitic acid), "soft" soaps result from the use of unsaturated fatty acids (e.g., linoleic acid, oleic acid) during formulation. In addition, although longer-chain fatty acids contribute to the stability of the foam, shorter-chain fatty acids (e.g., twelve carbons) allow for fast foam formation associated with small bubbles. Shaving soaps were first commercially manufactured in the 1830s and often were a mixture of tallow, stearic acid, potash, and caustic soda formed into a bar. Toward the 1940s, most shaving soaps were formulated to be used with a brush and mug. While some modern shaving soap preparations are nonlathering (only providing lubrication and moisture), foaming shaving soaps are typically designed so that mechanical agitation with the hands or use of the brush-and-mug method creates a foam that is used to coat the skin and provide efficient hydration. Standard modern foaming shaving soaps consist of fatty acids (e.g., stearic acid, palmitic acid, potassium myristate, sodium myristate), oils (e.g., coconut oil, palm kernel oil, corn oil), emollients (e.g., glycerin), and humectants (e.g., purified water), frequently formulated with alkaline chemical neutralizers (e.g., sodium hydroxide, potassium hydroxide), fragrance, and preservatives.

In the late 1940s, commercialization of aerosol shaving products led to the development of a variety of chemical formulations beyond a soap and water base. Aerosol foams consist typically of a potassium or sodium salt fatty acid mixture within a liquid soap base that foams upon application. Included are additional materials such as skin cleansers and antiseptics (e.g., triclosan, sodium lauryl sulfate), product thickeners (e.g., acrylate copolymers, cellulose polymer, carrageenan, hydroxyethylcellulose), surfactant/fatty acid neutralizers (e.g., triethanolamine), lubricants and emollients (e.g., lanolin, allantoin, sorbitol), hair and skin softeners/hydrating humectants (e.g., propylene glycol, glycerin, PEG compounds), and skin conditioners (e.g., *Aloe vera* gel, sodium cocoyl isothionate, vitamin E acetate, panthenol). In addition, herbal extracts (e.g., lime, lady's mantle, soapwort, peppermint, horsetail, watercress, lemongrass, sage, golden seal, orange flower), preservatives (e.g., BHT, DMDM hydantoin, quaternium-15, methylparaben, propylparaben, sodium benzoate), coloring pigments, and fragrance are often added to improve product quality and purity. Overall, these ingredients are combined in three-piece metal cans and applied to the skin with the use of propellants that include isobutane, propane, pentane, butane, and isopentane.

Specialized postfoaming shave gels, first developed in the early 1970s, frequently possess soap bases similar to traditional aerosol shaving formulations but also contain a polymer and other surfactant materials in the presence of selected hydrocarbons, which allows for the creation of a clear gel structure. Common ingredients in these products include deionized water, a fatty acid (palmitic acid)/triethanolamine salt mixture, fatty acid esters, a pentane/isobutane postfoaming mix, and a cellulose polymer. Mechanical movement of the gel over the skin surface creates a dense foam, providing additional lubrication and protection preferred by many consumers over traditional shaving foams.

SUNLESS TANNING PRODUCTS

While Greco-Roman, Elizabethan, and early American colonial societies viewed individuals with pale skin as belonging to the upper classes, in the twentieth century society began accepting bronzed skin. The suntan became the symbol of wealth and leisure, and brown- and beige-tinted powders and creams were created and brushed on the skin to supplement natural sun tanning. Although there is currently an increasing awareness of the detrimental effects of solar irradiation on human skin, a tanned appearance remains in fashion.

To achieve a darker tanned appearance without sun exposure, various sunless tanning formulations have become available over the counter. Most self-tanning products (e.g., lotions, foams, and sprays) contain the tanning agent dihydroxyacetone (DHA; $C_3H_6O_3$) as an active ingredient. Discovered as a skin darkener in the 1920s and first marketed as a tanning product ingredient in the 1960s, this carbohydrate can be derived from a vegetable source such as beets or sugarcane or produced commercially from glycerine using a bacterium that converts alcohol into acetic acid. The sunless tan created by applying the FDA-approved DHA cosmetic ingredient to the skin is artificial because no melanin is involved in the process. The action of DHA is limited to the upper layer of the epidermis (stratum corneum) and involves a reaction between DHA and skin proteins. In general, the Maillard browning reaction (also known as "nonenzymatic browning") between carbohydrates and amines is part of an extensive series of reactions that is the basis for the brown color caused by DHA. While the initial stages of the reaction are quite complex, the chemical reaction between DHA and amine groups of specific protein amino acids (e.g., arginine, lysine, and histidine) within the stratum corneum eventually yields brown polymers collectively known as melanoidins, large cross-linked molecules. The development of brown color usually requires two to four hours after application, and the color intensity is dependent on the formulation concentration of DHA in the product. Melanoidins are quite stable, so the sunless tan usually lasts between five and seven days from the initial application. However, the epidermis is quite regenerative and is completely replaced every thirty-five to forty-five days, so product reapplication is recommended to maintain the tanning effect. When used alone, DHA can cause dryness and create an uneven, orange/yellow, unnatural-appearing tan. Some products contain erythrulose, a natural ketose obtained by biofermentation, that is often applied in an emulsion in combination with DHA and leads to a more uniform, streakless tanning of the skin and longer-lasting tanning color. Tanning color tonality may also be affected by adding red coloring agents to the sunless tanning product, including iron oxides, D&C pigments, caramel, and carmine. In addition, other products added to sunless tanning products include water, emollients (e.g., cetyl alcohol, cyclopentasiloxane, cetyl palmitate, mineral oil, palm kernel oil, C12-15 alkyl benzoate), emulsifiers (e.g., cetyl hydroxyethylcellulose, polysorbate 60, steareth-20, inositol, polyquaternium-37), humectants (e.g., propylene glycol, glycerin, sorbitol), moisturizers (e.g., lanolin, cocoa butter, herbal extracts, lactic acid, xanthan gum, methylpropanediol), vitamins (e.g., tocopheryl acetate [vitamin E acetate], panthenol [provitamin A], tocopherol [vitamin E], panthenol [provitamin B5]), lubricants (e.g., dimethicone copolyol), texturizers and fillers (e.g., mica, glyceryl stearate, magnesium aluminum silicate), solvents (e.g., benzyl alcohol, stearyl alcohol), buffers (e.g., disodium EDTA), fragrance, alkalizers (e.g., sodium citrate, sodium hydroxide), and preservatives (e.g., BHT, sodium chloride, sodium metabisulfite, methylparaben, propylparaben, butylparaben). Active ingredients such as octocrylene, octyl methoxycinnamate, oxybenzone, or lawsone (the active component of henna; provides red-orange tint) may be added to provide sunscreen protection, since sunless tanning products alone do not provide skin protection against UV radiation similar to a melanin-based tan.

Although cosmetic bronzers can also yield a sunless tan with relatively immediate results (usually within forty-five to sixty minutes), these products, in the form of powders, creams, foams, and lotions, are essentially a form of brown-tinted makeup that can easily be removed with soap and water.

SUNSCREENS

In 7500 BC, Egyptian shepherds and hunters in the Nile valley used oil crushed from castor beans to protect their skin from sun exposure. An Australian chemist experimented with sunburn cream formulations in the 1930s, and a French chemist invented the first sunscreen in 1936. By the 1940s, a Florida scientist developed a suntan cream formulation, and this basic chemical composition is still marketed by a major manufacturer today.

The skin is an organ basically composed of two general tissue layers, the inner dermis and outer epidermis. Of the four different cell layers of the epidermis, it is the outermost, nonliving keratinized layer called the stratum corneum that provides a protective physical barrier against environmental factors, including UV radiation. UV light is irradiated from the sun to the surface of the earth in the wavelength range from approximately 290 nanometers. This UV wavelength range includes two different types of UV, UV-B (290 to 320 nanometers) and UV-A (320 to 400 nanometers). Although UV-A is capable of penetrating into the dermis, UV-B only penetrates into the epidermis. Both of these UV types may elicit skin changes, but the shorter-wavelength UV-B is significantly more energetic than UV-A at inducing these changes, including vitamin D production, erythema (sunburn), melanogenesis (melanin production), DNA damage, and some types of cancer (e.g., melanoma and squamous cell carcinoma). The less energetic longer-wavelength zone of UV-A radiation tends to elicit melanogenesis slowly but also may enhance the tissue-destructive effect of UV-B radiation.

Tanning is a process that occurs within the skin and involves the production of a black-brown pigment polymer called melanin. Produced by cells called melanocytes within the epidermis, light-colored unoxidized melanin granules are changed to their dark brown-colored oxidized form after exposure to UV light. This dark melanin then physically prevents the deeper layers of skin from sustaining damage from further UV exposure. A more sustained melanin-based tan may be achieved via a secondary tanning stage, which involves the production of new melanin pigment from the precursor amino acid tyrosine. Subsequent UV exposure may thus cause further melanin production and lengthening of the melanin polymer chains to produce darker skin coloration.

Commercial suntan and sunscreen products act by blocking UV-B selectively, allowing UV-A through to produce a slow tan, or by blocking both UV-B and UV-A and thus shielding the skin from the entire UV spectrum that reaches the earth's surface. Sunscreen products are manufactured to be chemically and photochemically stable, because the UV absorption capabilities of organic sunscreen active ingredients change over time. Overall, various inorganic and organic active UV-absorbing chemicals provide the sunscreen (or skin) protection factor (SPF) product rating. SPF is a ratio based on the skin-burning potential of the consumer with the sunscreen applied relative to lacking any sunscreen within a specific duration of sun exposure. For example, a sunscreen applied with an SPF of 6 indicates that the consumer may remain exposed to the sun approximately six times longer than if a sunscreen was not applied and achieve the same effect on the skin. Generally, higher SPF values tend to protect the skin of the consumer better because SPF is a measure of sunscreen product efficacy. Formulas must be soluble in a cosmetic base but water and sweat resistant, since active sunscreen ingredients must maintain a high concentration in the stratum corneum for several hours to ensure product goals. For many years, the active ingredient in many sunscreen lotions and creams has been *para*-aminobenzoic acid (PABA) or one of its esters. A UV-B protector, PABA inhibits an optimum ratio of quick tanning transmission to sunburn protection, but also has been shown to elicit an allergic response in many consumers. As an alternative, other UV-B protection active sunscreen ingredients are often used, including octyl methoxycinnamate, octyl salicylate, and homosalate (salicylic acid 3,3,5-trimethylcyclohexyl ester). UV-A protection active ingredients used include oxybenzone (2-hydroxy-4-methoxybenzophenone), dioxybenzone, or menthyl anthranilate. Inorganic metal oxides such as titanium dioxide and zinc oxide are commonly used as nonallergenic sunscreen active alternatives, as they both display absorbing properties throughout much of the UV-A and UV-B wavelength spectrum.

As the quality of the spreading agent is vital to consumer application ability (and thus protection potential), the agreeable cosmetic emulsion or solution should spread to form a continuous, coherent, and stable film on the skin. Thus, solvents (e.g., water, C12-15 alkyl benzoate, butyl octyl salicylate, isononyl isononanoate, alkyl salicylate, octyl dodecyl benzoate, propylene glycol benzoate), emulsifiers (e.g., cetearyl alcohol, ceteareth-20, dimethicone copolyol, propylene glycol, polysorbate 60, PEG 40 stearate), agents imparting water resistance (e.g., acrylates, cetyl dimethicone, PVP eicosane copolymer, maleated soybean oil), preservatives (e.g., disodium EDTA, imidazolidinyl urea), and fragrance are often added.

TOOTHPASTE

The ingredients of toothpaste, when used in tandem with tooth brushing, aid in oral hygiene goals such as plaque removal, decay resistance, promotion of remineralization, tooth cleansing and polishing, tooth stain removal, and breath freshening. The solid phase of the paste (a blend of agents) is typically suspended in a polyalcohol (e.g., aqueous glycerol or sorbitol) via a suspending agent. Also called dentifrice, toothpaste is composed of the following cleaning ingredients (in approximate volume percentages):

humectant and water, 40 to 70
buffers/salts/tartar control, 0.5 to 10
organic thickeners, 0 to 12
abrasives, 10 to 50
actives (e.g., triclosan), 0.2 to 1.5
surfactants, 0.5 to 2
flavor and sweeteners, 0.8 to 1.5
fluoride, 0.24

The two major ingredients are the detergents and abrasives. Detergents create the foaming action we associate with brushing with toothpaste, aiding in the retention of toothpaste within the mouth. Sodium lauryl sulfate is a detergent used in nearly all toothpastes to clean the surface of the teeth by acting as such a foaming agent. Detergents loosen food particles and other debris and aid in the movement of the abrasive across the gums and teeth surface. An abrasive typically consists of hard, very fine insoluble particles that act as scrubbing agents to remove stains and plaque as well as polish teeth. Nearly all modern toothpaste contains mildly abrasive hydrated silica particles (e.g., orthosilicic acid and disilicic acid). Other abrasive agents incorporated include calcium carbonate, dicalcium phosphate, and alumina trihydrate. The level of abrasivity is of paramount concern because the toothpaste must be abrasive enough to help remove stains and dental plaque without damaging tooth surfaces. Humectants provide toothpaste texture as well as moisture retention properties. Glycerin, sorbitol, and water are common humectants. Xylitol is an uncommon but superior humectant that has also been shown to enhance the anticavity action of fluoride. Thickeners also help create the texture of toothpaste and include agents such as carrageenan, cellulose gum, and xanthan gum. Flavoring agents and sweeteners (e.g., sodium saccharin) improve the taste of toothpaste, while coloring agents (e.g., titanium dioxide, white color) provide toothpaste with agreeable colors. In addition, toothpaste may also contain preservatives (e.g., sodium benzoate, methylparaben, ethylparaben) that prevent the growth of microorganisms, thus eliminating the need for refrigeration.

The development of toothpaste began as long ago as 300/500 BC in ancient countries of China and India. During the years 3000/5000 BC, Egyptians made toothpaste from powdered ashes of hooves of oxen, myrrh, powdered and burned eggshells, and pumice. In AD 1000, Persians added burnt shells of snails and oysters along with gypsum. In eighteenth century England, a tooth-cleansing "powder" containing borax was marketed in ceramic pots. By World War II, toothpaste was marketed in laminated tubes. In the 1950s, the scientific research findings that fluoride treatment led to dramatic reductions in dental cavities resulted in the widespread introduction of fluoride into toothpaste.

Many toothpastes are currently fluoridated with three compounds (e.g., sodium fluoride, sodium monofluorophosphate, and stannous fluoride) to give greater protection against tooth decay. Saliva serves as a reservoir for calcium and phosphate, which along with fluoride remineralizes tooth enamel. The human mouth may contain more than 500 types of microorganisms, including aerobic and anaerobic bacteria and fungi. Nearly immediately after a tooth cleansing, the teeth may be coated with a thin film (pellicle) derived mainly from saliva containing polysaccharides, fats, and proteins. The pellicle may be colonized by bacteria, and this forms a gel-like substance called plaque. The plaque provides a scaffold for bacteria to grow and metabolize carbohydrates, producing organic acids that may demineralize tooth enamel and underlying dentin, eventually forming dental caries (cavities). Normal tooth enamel contains hydroxyapatite $[Ca_{10}(PO_4)_6(OH)_2]$. An exchange reaction occurs in the presence of fluoride ion to form fluoroapatite $[Ca_{10}(PO_4)_6F_2]$, which is much more resistant to enamel-destroying acids produced by carbohydrate-metabolizing bacteria.

Some toothpastes contain ingredients (e.g., triclosan, xylitol, and stannous fluoride) that act as antimicrobial agents, chemically hindering the growth of plaque and gingivitis (gum inflammation)-causing bacteria. Recent improvements in toothpastes involve tartar (calculus) control. Tartar (mainly calcium phosphate $[Ca_3(PO_4)_2 \cdot H_2O]$) deposits on teeth as plaque hardens and can only be removed mechanically by a dental professional. Thus, tartar control toothpastes containing such active ingredients as sodium or potassium pyrophosphate $(Na_4P_2O_7)$ only slow the formation of tartar by inhibiting particular bacterial enzymes.

Baking soda (sodium bicarbonate $[NaHCO_3]$) is often added to toothpastes to act as a mild abrasive and to neutralize oral acids originating from the decomposition of food materials. Some desensitizing toothpastes may work to minimize the pain of tooth hypersensitivity when hot or cold foods are ingested. Two effective ingredients in treating sensitive teeth and gums are strontium chloride $(SrCl_2)$ and potassium nitrate (KNO_3), which block the sensation of pain via nerve transmission from such teeth to the central nervous system.

3

Health and Medical Products

ALLERGY TREATMENTS (ANTIHISTAMINES)

Allergy (also called hypersensitivity) involves an inappropriate or excessive immune response of the body to foreign substances called allergens. Allergens are simply antigens (antibody-generating substances) that trigger allergic reactions. Allergies are caused by an immune response to a normally nonharmful substance (e.g., pollen, dust, mold spores, animal dander) that comes in contact with agranular white blood cells (lymphocytes) specific for that substance, or antigen. An allergic reaction is an immune response that should not occur, as the substance that usually triggers the response is inherently not a danger to the body.

An antibody (also called an immunoglobulin [Ig]) is a soluble protein secreted by specialized blood cells (e.g., B-lymphocytes and plasma cells) in response to the body being exposed to an antigen. Although body fluid antibodies are formed in response to a vast number of different antigens, all antibodies are grouped into five classes: IgG, IgD, IgM, IgA, and IgE. Antibodies are capable of binding specifically with an antigen, and when an antibody molecule binds to its corresponding antigen, an antigen-antibody complex is formed. Allergic sensitivity to an allergen is mediated through IgE, such that when the body is first exposed to an antigen, cells called macrophages engulf the antigen (through a process called phagocytosis) and the antigen is subsequently presented on the macrophage combined with glycoproteins on the cell membrane surface. Cells called T-lymphocytes then interact with the antigen to initiate a T-cell-mediated response; additionally, B-lymphocytes produce IgE to complex with the antigen. The plasma membranes of mast cells and cells called basophils possess high-affinity IgE-binding sites. Basophils (granular white blood cells) and mast cells (normally found in loose connective tissue) contain secretory granules that store a variety of inflammatory mediator chemicals, including histamine. Once the IgE is fixed to mast cells and basophils after the primary exposure, a secondary exposure to the same allergen causes histamine to be released from the secretory granules. The allergen binds to the IgE molecules on the mast cell surface, causes cross-linking (also called cross-bridging) of the IgE molecules, which then leads to mast cell "degranulation," whereby the histamine-containing granules are released into the surrounding tissues. Release of histamine subsequently causes the release of cytokines and numerous other chemical mediators of inflammation. Inflammation is the reaction of living tissue to injury, infection, or irritation. Inflamed tissues are characterized by pain, swelling, redness, and heat.

Histamine [2-(4-imidazolyl)-ethyl-amine], a vasoactive monoamine formed by the decarboxylation of histidine by the enzyme histidine carboxylase, chemically mediates local immune responses. It is located in most body tissues but is highly concentrated in the lungs, skin, and gastrointestinal tract. In the central nervous system (CNS), it may serve as a neurotransmitter. Histamine acts by binding to receptors on target cells, with different cell types expressing different currently characterized histamine receptor (H) types (e.g., H_1, H_2, H_3, H_4). Overall, histamine contributes to the inflammatory response, acts on the smooth muscle of blood vessels to cause vasodilation (increase in blood vessel diameter), causes an increase in the permeability of blood vessel walls, and affects nearby sensory nerves, resulting in itching (also called pruritus). Depending on the route of allergen exposure (e.g., inhalation, skin or eye contact, ingestion, etc.), the effects of histamine cause the familiar symptoms of allergy, including sneezing, inflammation of the nasal passageways, nasal itching, watery nasal discharge, and itchy, inflamed, tearing eyes. When released in the lungs, histamine causes smooth muscle contraction of the airway bronchioles, which is an attempt of the body to prevent the offensive allergens from entering the lung tissue. Unfortunately, this type of response leads to the symptoms of wheezing and shortness of breath similar to that experienced by individuals with the life-threatening condition called asthma.

Allergies are frequently treated by drugs called antihistamines, because they antagonize/inhibit the activity of histamine. The use of antihistamines allows individuals to live more safely and comfortably by counteracting immunological mistakes and alleviating the annoying and potentially harmful symptoms associated with allergens. In 1937, D. Bovet and A. Staub discovered the first H_1 receptor antagonist. This discovery marked the "first generation" of antihistamines characterized to treat allergic diseases, and modern over-the-counter oral antihistamines continue to be a reasonable and effective treatment for the various symptoms associated with allergies. These drugs are efficiently absorbed into the bloodstream from the gastrointestinal tract, metabolized primarily within the liver, excreted through the urine, and usually possess a four- to six-hour duration of action. Antihistamines suppress the wheal (swelling) and flare (vasodilation) response after exposure to allergens by blocking the binding of histamine to its receptors on nerves, vascular smooth muscle, glandular cells, endothelium tissue, and mast cells. Classic first-generation antihistamines block the action of histamine at specific histamine receptors (e.g., H_1 receptors). First-generation antihistamines are small lipophilic molecules, so they may cause adverse effects (e.g., sedation, impaired cognition, blurred vision, gastrointestinal symptoms, dryness of mouth, heart palpitations, urinary retention) because their structure closely resembles that of blockers of cholinergic (specifically muscarinic) and α-adrenergic receptors of the autonomic nervous system and because of their ability to cross the blood-brain barrier to affect the CNS. These H_1 receptor antagonists are reversible, competitive inhibitors of the pharmacological actions of histamine on H_1 receptors. Dozens of first-generation H_1-blocking compounds are available in over-the-counter oral allergy treatments and may include chemical classes of drugs structurally related to histamine, such as ethanolamines, including diphenhydramine (2-diphenylmethoxy-N,N-dimethylethanamine; $C_{17}H_{21}NO$), doxylamine (N,N-dimethyl-2-[1-phenyl-1-(2-pyridinyl) ethoxy] ethanamine; $C_{17}H_{22}N_2O$), and clemastine ([R-(R^*, R^*)]-2-[2-[1-(4-chlorophenyl)-1-phenylethoxy] ethyl]-1-methylpyrrolidine; $C_{21}H_{26}ClNO$), and alkylamines (also known as propylamines), including brompheniramine (2-[p-bromo-α-(2-dimethylaminoethyl) benzyl]pyridine; $C_{16}H_{19}BrN_2$) and chlorpheniramine (2-[p-chloro-α-(2-dimethylaminoethyl)benzyl]pyridine; $C_{16}H_{19}ClN_2$).

Most antihistamines do not chemically inactivate or physiologically antagonize histamine, nor do they prevent histamine release. However, loratadine [4-(8-chloro-5,6-dihydro-11H-benzo[5,6]cyclohepta[1,2-b]pyridin-11-ylidene)-1-piperidinecarboxylic acid ethyl ester; $C_{22}H_{23}ClN_2O_2$], a long-acting (often twenty-four hours) second-generation piperidine-derivative antihistamine with selective peripheral H_1 receptor antagonist activity, is suspected of additionally decreasing histamine release from basophils. Second-generation (also called nonsedating) antihistamines are more lipophobic than first-generation antihistamines and are thought to lack CNS and cholinergic receptor-blocking effects when used at therapeutic doses.

Antihistamines are found in combination with other ingredients (e.g., decongestants, analgesics) in many over-the-counter cold, sinus, and allergy medications. In addition, liquid products often contain large amounts of alcohol as well.

ANALGESICS

An analgesic is a drug that relieves the sensation of pain. Pain is often described as a primitive physiological experience that, through an atmosphere of suffering, acts as a protective mechanism to warn of actual or potential damage to biological tissues in humans and nearly all other types of animals. The principle nociceptors (pain receptors) include several million free sensory nerve endings with large receptive fields that are present in all of the tissues and organs of the body (excluding the brain). While a small number of nociceptors are located in deep tissues and visceral organs, they are especially common in areas such as the superficial portions of the skin, joint capsules, bone tissue, and surrounding blood vessel walls. Nociceptors are often categorized as belonging to one of three different population types: those sensitive to temperature extremes, to mechanical damage (e.g., swelling/inflammation), or to the presence of chemicals/toxins released by damaged and/or injured cells.

Stimulation of the dendrites of the nociceptor neurons by a specific stimulus (e.g., temperature change, mechanical damage, toxin exposure) often causes a neurophysiological change (called a depolarization), which may then lead to the transmission of a nerve impulse (in the form of an action potential) to the CNS. Two types of nerve cell fibers (called axons), type A and type C, carry painful sensations. Type A, myelinated fibers carry sensations of "fast pain" (e.g., prickling pain), and type C, unmyelinated fibers carry sensations of "slow pain" (e.g., burning, itching, and aching pain). Type A and type C fibers transmit nerve impulses through neurons of the lateral spinothalamic pathway of the spinal cord and brain, eventually transmitting pain impulses through the thalamus and to the primary sensory cerebral cortex (within the cerebrum) of the brain, where conscious attention of the pain is processed.

After tissue injury, damaged cell membranes release an omega six fatty acid derivative compound called arachidonic acid into the interstitial fluid space surrounding tissue cells. Within the interstitial fluid, an enzyme called fatty acid cyclo-oxygenase (also called prostaglandin endoperoxide synthetase) converts arachidonic acid molecules to prostaglandins (PGs). There are currently two known forms of cyclo-oxygenase, cyclo-oxygenase-1 (COX-1) and cyclo-oxygenase-2 (COX-2). The COX-1 isoform is constitutively expressed in most normal cells and tissues, including the stomach. COX-2 is generally induced only in settings of inflammation by cytokines and other chemical mediators of the inflammatory response. However, COX-2 is constitutively expressed in certain areas of the kidney and brain. In general, COX-1 converts arachidonic acid to PGI_2, which protects the stomach lining from acidic digestive juices, whereas COX-2 converts arachidonic acid to PGE_2, which acts on nerve endings to cause the sensation of pain. Thus, the enzymatic processes involved with PG synthesis are of pharmacological concern, because PGs, named for the seminal fluid and prostate tissue in which they were discovered, are chemical messengers of the immune system responsible for inflammation and pain.

The capacity of PGs to sensitize pain receptors to mechanical and chemical stimulation appears to result from a lowering of the depolarization threshold of the nociceptors of type C fibers, thereby increasing the nerve transmission of pain sensation to the brain. Other potent pain-producing nociceptor-activating chemicals include bradykinin and cytokines. Bradykinin, a member of the kinin system group of proteins, is derived by cleavage of precursor plasma protein (plasma kininogen) molecules and is a potent vasodilator, a contractor of a variety of different kinds of extravascular smooth muscle tissue (e.g., bronchial), and an inducer of increased vascular permeability. Cytokines are low-molecular-weight peptides or glycoproteins produced by multiple cell types, such as lymphocytes, monocytes/macrophages, mast cells, eosinophils, and endothelial cells lining blood vessels. The four major categories of cytokines are interferons, colony-stimulating factors, tumor necrosis factors, and interleukins. Interferons interfere with virus replication; colony-stimulating factors support the growth and differentiation of cellular elements within bone marrow; tumor necrosis factors (TNFs) cause a bleeding (hemorrhagic) tissue death (necrosis) of tumors when injected into various types of animals; and interleukins (ILs) allow for communication between (inter-) various populations of white blood cells (leukocyte-leukin). Bradykinin and cytokines (e.g., TNFa, IL-1, and IL-8) liberate PGs to promote enhanced pain sensitivity (called hyperalgesia).

Analgesic drugs are a heterogeneous group of compounds, often chemically unrelated, that share certain therapeutic actions and side effects. Over-the-counter (OTC) oral analgesics contain active ingredients including acetaminophen and/or nonsteroidal anti-inflammatory drugs (NSAIDs), such as aspirin, ibuprofen, ketoprofen, and naproxen sodium. OTC oral analgesics are generally recommended safe for treatment of pain for approximately seven to ten days, with labels warning individuals to seek professional medical advice for chronic pain symptoms. Whereas the general side effects of all OTC oral analgesics may include kidney damage, the specific side effects of oral NSAIDs may include irritation of the inner stomach lining, general digestive upset, ulcers, and bleeding in the digestive tract. In general, acetaminophen and NSAIDs are nonselective COX enzyme inhibitors (they inhibit both COX-1 and COX-2 enzymes with little selectivity) and thereby decrease the sensation of pain by inhibiting PG production.

Acetaminophen (paracetamol; *N*-acetyl-*p*-aminophenol; $C_8H_9NO_2$) is the least toxic member of a class of analgesic (and antipyretic [antifever]) medications known as the *p*-aminophenols. A major metabolite of phenacetin (the so-called coal tar analgesic) and acetanilide, acetaminophen is an effective and fast-acting analgesic that acts centrally to relieve mild to moderate pain. As early as 1886, acetanilide was used to alleviate fever, but it proved too toxic for general public use. Subsequent investigations of similar chemical compounds in the 1880s led to trials of *p*-aminophenol, which also proved to be excessively toxic for human use, and phenacetin (also called acetophenetidin; it was banned by the U.S. Food and Drug Administration in 1983 after reports of its tendency to cause human kidney damage and blood disorders when used excessively). In 1893, acetaminophen was introduced, and since 1949, when it was recognized as the major active metabolite of both acetanilide and phenacetin, it has gained a prominent household position as an effective alternative to NSAIDs as an analgesic-antipyretic. While promoted in 1955 for children's fever and pain, acetaminophen became available as an OTC drug in 1960. However, unlike aspirin, acetaminophen possesses weak anti-inflammatory activity; thus, it is not a useful agent in treating pain associated with inflammation. Acetaminophen acts to alleviate the pain of fever by effectively inhibiting the cyclo-oxygenase enzyme in the brain and PG production, producing its analgesic effect by increasing the pain threshold. The failure of acetaminophen to exert anti-inflammatory activity at sites of inflammation in peripheral body tissues may be attributed to the fact that acetaminophen is only a weak inhibitor of the enzyme cyclo-oxygenase. Moreover, acetaminophen is thought to inhibit this enzyme activity only in environments that are low in the chemical called peroxide (e.g., brain tissue). This characteristic may in part explain the poor anti-inflammatory activity of acetaminophen, as sites of peripheral inflammation usually contain increased concentrations of peroxides released by white blood cells (leukocytes). Primarily metabolized within the liver, minor metabolites contribute significantly to the toxic effects of acetaminophen. These include potentially lethal and/or irreversible liver damage when consumed in excess of recommended therapeutic doses. However, single or repeated therapeutic doses of acetaminophen typically elicit no deleterious effects on the cardiovascular and respiratory systems. In addition, unlike aspirin and other salicylate-based analgesic drugs, acetaminophen at normal therapeutic doses does not affect acid-base balance within the body, does not interfere with bleeding time (hemostasis) or with kidney tubular secretion of uric acid wastes, does not inhibit platelet aggregation, and does not produce stomach irritation, erosion, or bleeding after administration.

The OTC oral NSAIDs, including aspirin, ibuprofen, ketoprofen, and naproxen sodium, are particularly effective in settings in which inflammation has caused the sensitization of pain receptors to normally painless mechanical or chemical stimuli. It is generally accepted that pain associated with inflammation and tissue injury most likely results from local stimulation of nociceptors and excessive pain sensitivity. Pain after inflammation can be attributable to damage to the actual nerves, pressure from the swelling of excess tissue fluid buildup on the nerve endings, or irritation from cellular chemical toxins released from injured tissues. During the inflammatory process, PGs are secreted into the interstitial space between tissues and can prolong and increase the severity of the inflammation response and associated pain. Thus, NSAIDs act specifically as anti-inflammatory agents, and thereby decrease the sensation of pain, by inhibiting COX activity and PG synthesis. The vast majority of NSAIDs available as OTC oral analgesics are organic acids and, in contrast to aspirin, act as reversible, competitive inhibitors of cyclo-oxygenase activity. Aspirin is unique among the NSAIDs in that it chemically modifies both the COX-1 and COX-2 enzymes, which results in an irreversible inhibition of cyclo-oxygenase enzyme activity. Because aspirin and other NSAIDs are organic acids, they are well absorbed orally, bind to plasma blood proteins, accumulate at sites of inflammation, and thus act effectively as anti-inflammatory drugs. In contrast to aspirin, whose duration of pain-alleviating action is determined by the rate of new cyclo-oxygenase enzyme synthesis, the duration of action of all other NSAIDs (reversible inhibitors of cyclo-oxygenase activity) is predominantly related to the ability of the body to clear the NSAID drug through the kidneys via urine excretion.

Aspirin [2-(acetyloxy)benzoic acid; acetylsalicylic acid; $C_9H_8O_4$] is a member of a family of chemicals called salicylates, which have been used to treat a variety of conditions for more than 2,500 years. The Latin term for willow, *salix*, provides the historical basis of the name of this family, whose molecules resemble those of both the alcohol and the acid forms of salicylate. The medicinally therapeutic effects of willow tree (genus *Salix*) bark were known to many diverse cultures of people for centuries. As early as 400 BC, the Greek physician Hippocrates described a bitter powder extracted from willow bark and closely related plants and recommended its use to ease pain and reduce fevers. The use of this product was also promoted by Galen, a second-century Roman physician, and mentioned in medical texts of the Middle Ages and Renaissance. In 1757, Reverend Edward Stone of England wrote about the success of willow bark in the cure of fevers and aches in a letter to the president of the Royal Society. In 1829, a pharmacist known as H. Leroux demonstrated that a bitter glycoside called salicin was the active antipyretic ingredient in the willow bark. In 1839, the Italian chemist R. Piria hydrolyzed salicin into glucose and salicyl alcohol and subsequently oxidized salicyl alcohol to salicylic acid. In 1853, C. von Gerhardt obtained salicylic acid from the reaction of salicylaldehyde with strong base. This fragrant aldehyde was itself isolated from meadowsweet flowers, which belong to the genus *Spiraea*. Salicylic acid was also synthesized by a process discovered by H. Kolbe and E. Lautemann in 1860, which led to the introduction of salicylic acid, and related compounds such as sodium salicylate, in 1875 for the treatment of fever and arthritis. Although toxic to the stomach and often causing diarrhea and vomiting, salicylic acid and related compounds were used at high doses to treat pain and inflammation in diseases such as arthritis and to treat fever in illnesses such as influenza (flu) for many years. In 1897, the overwhelming success of the salicylate drugs, and also the desire to provide relief for his father's rheumatoid arthritis, prompted F. Hoffman, a German chemist employed by Friedrich Bayer & Co., to find a less toxic alternative analgesic. He prepared acetylsalicylic acid by chemically modifying salicylic acid (through a reaction with acetic anhydride) based on the work of C. von Gerhardt more than 40 years earlier and demonstrated that in its acetylated form the salicylate was easily tolerated and possessed a potent analgesic effect. After a clear demonstration of its anti-inflammatory and analgesic effects, H. Dreser, a German chemist also employed by Friedrich Bayer & Co., introduced acetylsalicylic acid into medicine in 1899 in the form of orally administered capsules and powder in envelopes. With commercial production of the acetylated salicylic acid, the term "aspirin" was introduced; it was derived from adding an "a" (for acetylated) to a portion of an older name of the acid, spiraeic acid, which was derived from the genus *Spiraea* for meadowsweet. Since its acceptance into medical practice at the beginning of the twentieth century, aspirin has become one of the most widely available and used drugs for the treatment of illness or injury.

Aspirin covalently modifies both COX-1 and COX-2 enzymes, resulting in an irreversible inhibition of COX enzyme activity. In the structure of COX-1, aspirin acetylates the serine amino acid residue at position 530 of the long interior protein channel of the enzyme. This chemical alteration prevents the binding of arachidonic acid to the active site of the enzyme within the interior enzyme core, prohibiting the enzyme from catalyzing the chemical transformation of arachidonic acid to pain-promoting PGs. In COX-2, aspirin acetylates a homologous serine amino acid residue at position 516 of the long interior protein channel of the enzyme. Thus, covalent modification of the COX-2 enzyme by aspirin blocks the PG synthesis activity of this isoform similar to COX-1. Besides acting as an analgesic, an antipyretic, and an anti-inflammatory drug, aspirin may also cause changes in kidney function and inhibit blood platelet function by irreversible inactivation of the cyclo-oxygenase enzyme

Most antacid products contain one or more of four active alkaline ingredients: aluminum salts [e.g., aluminum hydroxide; $Al(OH)_3$], magnesium salts [e.g., magnesium carbonate; $MgCO_3$ and magnesium hydroxide (milk of magnesia); $Mg(OH)_2$], calcium carbonate (chalk; limestone; $CaCO_3$), and sodium bicarbonate (baking soda; $NaHCO_3$). Several antacids are formulated with combinations of magnesium and aluminum salts [e.g., magaldrate; $AlMg(OH)_5$] to neutralize stomach acid, to decrease the action of pepsin (a digestive enzyme), and to prevent deleterious side effects, including constipation and diarrhea. In addition, many types of bismuth mineral salts (e.g., bismuth subsalicylate), which are multipurpose intestinal medicinal agents, are also used as antacids because they increase alkaline secretion to counteract any acid production in the stomach. As an antiulcer agent, bismuth subsalicylate also coats and protects irritated and inflamed gastrointestinal lumen tissue.

Effervescent antacids are those products that contain sodium bicarbonate and citric acid. These products give an effervescent action when dissolved in water because of the evolution of carbon dioxide (CO_2) gas when sodium bicarbonate and citric acid react in a hydrated environment. It has been suggested that the release of carbon dioxide by sodium bicarbonate during this reaction may provide relief from the discomfort of overeating by inducing belching, which aids in the expulsion of swallowed air. Sodium citrate, which is also produced in the reaction of sodium bicarbonate and citric acid, may also accept hydrogen ions (H^+) and revert back to citric acid. Thus, the sodium citrate, which is produced when effervescent antacids are dissolved in water, also acts as a stomach acid-neutralizing antacid.

Some antacids contain additional active ingredients, including analgesics (e.g., aspirin) to treat headaches and antigas/antibloating products (e.g., simethicone, magnesium trisilicate) to treat uncomfortable feelings often associated with heartburn as a result of overeating.

Histamine receptor-2 (H_2) blockers include chemicals such as cimetidine (N-cyano-N'-methyl-N''-[2-[[(5-methyl-1H-imidazol-4-yl)methyl]thio]ethyl]guanidine; $C_{10}H_{16}N_6S$), famotidine (3-[[[2-[(aminoiminomethyl)amino]-4-thiazolyl]methyl]thio]-N-(aminosulfonyl)propanimidamide; $C_8H_{15}N_7O_2S_3$), nizatidine (N-[2-[[[2-[(dimethylamino)methyl]-4-thiazolyl]methyl]thio]ethyl]-N'-methyl-2-nitro-1,1-ethenediamine; $C_{12}H_{21}N_5O_2S_2$), and ranitidine (N-[2-[[[5-[(dimethylamino)methyl]-2-furanyl]methyl]thio]ethyl]-N'-methyl-2-nitro-1,1-ethenediamine; $C_{13}H_{22}N_4O_3S$). These four individual drugs act as histamine antagonists (antihistamines) by competitively inhibiting the binding of histamine at histamine H_2 receptors. Such receptors are located in HCl-secreting parietal cells within the stomach and in the heart. Histamine (2-(4-imidazolyl)-ethyl-amine) is a chemical mediator produced by the decarboxylation of histidine by the

within the platelets, thereby inhibiting the natural ability of platelets to help form blood clots. As a nonselective COX enzyme inhibitor, aspirin also inhibits the biosynthesis of PGs within the stomach, which normally protect the gastrointestinal tract by inhibiting acid secretion in the stomach and promoting the secretion of protective mucus in the intestine. Thus, aspirin may render the stomach more susceptible to damage and has a strong tendency to cause harmful gastrointestinal side effects, including heartburn (acid reflux) and ulceration of the stomach and/or small intestine.

When aspirin reaches the stomach, the ester is converted back into salicylic acid, which then enters the bloodstream to relieve pain by interfering with PG synthesis. The rate at which the acetylsalicylic acid bound with a binder (an inert chemical-binding agent) in a solid orally administered aspirin tablet disintegrates in the stomach is dependent upon pH, such that as pH increases (more alkaline environment), the tablet breaks apart more easily, and the faster the "free" salicylic acid is available to be absorbed into the bloodstream. "Buffered" aspirins consist of a combination of aspirin and one or more bases, including magnesium carbonate ($MgCO_3$), magnesium hydroxide [$Mg(OH)_2$], aluminum hydroxide [$Al(OH)_3$], and aluminum glycinate (aluminum salt of the amino acid glycine). While many scientific studies have failed to produce any evidence that buffered aspirin results in quicker or increased overall analgesic effects compared with nonbuffered aspirin, the addition of these bases to the aspirin tablet has been shown to increase the rate of disintegration and subsequent drug absorption into the bloodstream.

Ibuprofen, ketoprofen, and naproxen sodium are all arylpropionic acid derivatives that represent a group of effective and useful analgesic, antipyretic, anti-inflammatory NSAIDs. All of these drugs interfere with the synthesis of prostaglandins via cyclo-oxygenase. For example, ibuprofen and naproxen function as anti-inflammatory drugs by physically plugging (not chemically altering) the long interior protein channels of COX enzymes, thereby preventing molecules of arachidonic acid from entering these enzyme channels and undergoing a chemical transformation in the enzyme's core into various types of PGs. Although all of these drugs are effective cyclo-oxygenase enzyme inhibitors, there is substantial variation in their potency. For example, naproxen is roughly twenty times more potent than aspirin, whereas ibuprofen, ketoprofen, and aspirin generally possess the same cyclo-oxygenase enzyme inhibition potency. Like aspirin, all of these agents may alter platelet function and bleeding time and cause stomach irritation and toxicity. In addition, naproxen in particular may also have prominent inhibitory effects on the functioning of white blood cells.

Reclassified by the Food and Drug Administration from a prescription to an OTC drug in 1984, ibuprofen (*p*-isobutylhydratropic acid; $C_{13}H_{18}O_2$) was the first member of the propionic acid class of NSAIDs to come into general public use. Ketoprofen (*m*-benzoylhydratropic acid; $C_{16}H_{14}O_3$) was reclassified similar to ibuprofen in 1995 and shares the pharmacological properties of other propionic acid-derivative NSAIDs. However, although ketoprofen may alleviate pain associated with inflammation by decreasing PG synthesis via inhibition of the cyclo-oxygenase enzyme, this drug may also alleviate inflammatory pain by stabilizing membranes of the cellular organelles called lysozymes, thereby preventing the release of toxic and inflammation-promoting chemicals into body tissues. Ketoprofen may also decrease pain and inflammation by counteracting the actions of the inflammation-promoting chemical bradykinin. Naproxen [d-2-(6-methoxy-2-naphthyl)propionic acid; sodium salt: $C_{14}H_{13}NaO_3$] was reclassified in 1994 and also inhibits the cyclo-oxygenase enzyme, thereby interfering with PG synthesis. Interestingly, the half-life drug presence of naproxen in the blood plasma after oral administration is approximately fourteen hours, and this value is increased about two-fold in elderly individuals. In comparison with naproxen, the half-life drug presence of ibuprofen and ketoprofen in the blood plasma after oral administration are each approximately two hours. Thus, for OTC oral naproxen (and naproxen sodium), dosing instructions caution individuals wishing to avoid the toxic physiological effects of the drug not to ingest in excess of three caplets (or tablets) in a twenty-four-hour period unless directed by a physician, and elderly individuals (over sixty-five years of age) are cautioned not to ingest in excess of one caplet every twelve hours unless directed by a physician.

ANTACIDS

The stomach, a J-shaped organ located between the esophagus and the small intestine within the gastrointestinal tract, performs many important body functions, including bulk storage of ingested food, mechanical breakdown of ingested food, production of a specialized glycoprotein called intrinsic factor that allows the body to absorb vitamin B_{12}, and disruption of chemical bonds within food material (e.g., proteins) through the action of acids and enzymes. The inner mucosal tissue layer of the stomach contains many deep gastric pits, which in turn lead into the gastric glands that collectively produce stomach secretions called gastric juice. Parietal cells, specialized secretory cells, are especially common along the proximal portions of each gastric gland. These cells secrete intrinsic factor and hydrochloric acid (HCl). The secretory activities (i.e., release of HCl) of parietal cells maintain the inner stomach contents at a relatively low pH between 1.5 and 2.0, whereas normal body extracellular fluid pH ranges between 7.35 and 7.45 (relatively neutral). The acidic pH of

gastric juices within the stomach allows for the destructi potentially harmful microorganisms within food, the dena teins within food, the inactivation of food enzymes, the ingested plant cell wall materials and meat connective tis proper activation and function of pepsin, a protein-dige secreted by other cells (chief cells) within the gastric glands

"Indigestion" refers to any number of gastrointestinal co can include excess gas production and upset/sour stom; burn" is a symptom; it refers to a burning sensation that is by the regurgitation of stomach acid from the stomach, up cardiac sphincter (the ring of smooth muscle that normally i movement of food from the esophagus to the stomach), into agus (the muscular tube that connects the stomach to the thi is positioned within the chest cavity. The sensation of heartb the result of a medical condition known as acid reflux; it g cludes symptoms such as a burning or discomfort behind the moves upward toward the throat and tends to worsen after ea lying down or bending over, a bitter or sour taste of acid at t the throat, and/or frequent burping or bloating. In some case cess acid production may lead to a condition called gastritis, tion of the underlying layers of stomach wall tissue. Persistent this tissue may promote the formation of peptic ulcers, erosic esophagus and/or stomach wall. A common predisposing facto formation is hypersecretion of HCl by the parietal cells lining t ach (also called hyperchlorhydria). Antacids/acid blockers are te used to treat hyperchlorhydria.

Over-the-counter oral antacids/acid blockers are a family of drugs that include antacids, histamine receptor-2 blockers, and pump inhibitors. Antacids help prevent damage to the inner muc sue layer of the gastrointestinal tract (e.g., stomach, esophagus) b ing in neutralizing excess highly acidic gastric juices. While minera antacids have been in world use since the third century, the first co cial stomach antacid seltzer was manufactured in the United St; Captain I. Emerson in 1891. Modern antacids do not completel; tralize stomach acid, as this would compromise digestive process(potentially cause "acid rebound," in which additional HCl is secre counteract such a dramatic change in stomach pH. However, these can increase the pH level within the stomach from approximately more alkaline level between 3 and 4. This increase neutralizes near percent of stomach acid and provides significant relief from minor h burn for many individuals. All of these products provide effective within a minute or less, and the duration of action is usually between minutes and two hours after treatment.

enzyme histidine decarboxylase. Within the gastric glands of the stomach, specialized enteroendocrine cells called histaminocytes release histamine, which binds to H_2 receptors and aids in the stimulation of HCl secretion by parietal cells. Oral H_2 receptor blockers typically require at least thirty minutes to one hour to initiate gastric acid secretion inhibition but then usually provide heartburn relief for up to six hours after treatment. In addition, these drugs are used in the management of peptic ulcers to allow for an increase in tissue healing and to prevent recurrences.

Proton pump inhibitors (PPIs) are a class of drugs that decrease the production of gastric acid by targeting structures located within the stomach parietal cells called proton pumps. Omeprazole (5-methoxy-2-[[(4-methoxy-3,5-dimethyl-2-pyridinyl)methyl]sulfinyl]-1H-benzimidazole; $C_{17}H_{19}N_3O_3S$) is an example of a chemically active ingredient found in oral over-the-counter PPI products. Histamine, the neurotransmitter acetylcholine, and the hormone gastrin all work through the proton pump to allow the parietal cells to produce HCl. This active transport (which requires energy in the form of adenosine triphosphate) proton pump moves hydrogen ions (H^+) produced within the parietal cells into the stomach lumen and pumps potassium ions (K^+) back into the parietal cell from the lumen. As hydrogen ions are secreted into the stomach lumen, a different active transport pump moves chloride ions (Cl^-) into the lumen to maintain an electrical equilibrium (balance) within the stomach. However, PPIs prevent the H^+/K^+ pump from functioning; thus, the movement of H^+ into the stomach lumen to form HCl is blocked. Overall, the onset of action of an oral PPI is approximately one hour after treatment. However, the duration of heartburn relief and inhibition of HCl production may extend for as long as sixteen hours.

ANTIDIARRHEALS

Diarrhea refers to increased water content in the stool compared with an individual's usual patterns. The frequency and volume of unformed stool are also increased. Food is propelled through the digestive tract by peristalsis, rhythmic contractions of muscles surrounding the digestive system. An increase in the overall rate of peristalsis along the digestive tube (called rapid gastrointestinal transit) can result in diarrhea, as there is insufficient time for water extraction from fecal material within the large intestine. Diarrhea may also be caused by osmotic attraction of water into the lumen of the gut (osmotic diarrhea), secretion of excess fluid into the gut (or decreased absorption; secretory diarrhea), and/or the exudation of fluid from an inflamed intestinal mucosal surface (exudative diarrhea).

Acute (short-term) diarrhea refers to a sudden change of bowel movements such that there is a frequent (three or more times per day) passage of voluminous loose stool for a period of less than several weeks. Sometimes associated with fever, malaise, dehydration, and/or abdominal cramping or pain, acute diarrhea usually lasts for a period of several days. Bacterial, parasitic, and/or viral infections are the most common cause of acute diarrhea and are sometimes manifested after the drinking or eating of toxin-contaminated water and/or food ("food poisoning"). Diarrhea often affects individuals who travel to developing countries ("traveler's diarrhea"), sometimes as a result of inadequate sanitation and contaminated food and water. The ingestion of numerous medications (e.g., antacids, antibiotics, antihypertensive medications, laxatives, magnesium supplements, potassium supplements, various cardiac medications) or of malabsorbed sugar found in many "sugarless" foods and drinks (e.g., fructose, lactose, mannitol, sorbitol) may also cause acute diarrhea. Chronic diarrhea refers to the physiological effects of acute diarrhea persisting for more than four weeks. Causes of chronic (long-term) diarrhea may include abdominal surgery, abdominal radiation treatments, chronic use of acute diarrhea-causing medications, various endocrine diseases, circulatory problems, neurological disease, chronic bacterial and/or parasitic infections, bowel disease, cancer of the digestive system, systemic chemotherapy cancer treatment, food allergy syndromes, and food maldigestion or malabsorption. Overall, diarrhea may lead to overall dehydration and loss of vital electrolytes, and prompt treatment is recommended to avoid serious health problems.

Treatment of various types of diarrhea may include the use of oral over-the-counter liquid, tablet, caplet, or chewable tablet antidiarrhea agents. These generally include constipating agents, absorbents, and antisecretory agents (e.g., bismuth salts). Constipating agents, such as opiates, affect the circular and longitudinal muscles of the intestine, increasing muscle tone while slowing the passage of stool through the intestine. This allows for increased water and salt reabsorption time, along with increased fecal consolidation and dehydration. Opiates may also elicit effects through central nervous system (brain and spinal cord) opioid receptors or by decreasing intestinal fluid secretion. These agents include diphenoxylate and loperamide, both structurally related to 4-phenylpiperidine opioid analgesics (pain relievers), but have negligible opiate effects and a low abuse potential.

Absorbents, such as kaolin (a hydrated aluminum silicate clay mineral), pectin (a complex carbohydrate often extracted from fruits), and attapulgite (a hydrated magnesium aluminum silicate clay mineral), form a thickening powder that absorbs excess fluid and any bacterial toxins present (action of kaolin and attapulgite) and increases the consistency of the stool by forming a viscous colloidal solution (action of pectin). However, these agents may also absorb essential body enzymes and nutrients.

Many types of bismuth mineral salts (e.g., bismuth subsalicylate) are generally multipurpose intestinal medicinal agents. As an antisecretory agent, bismuth subsalicylate coats and protects irritated and inflamed intestinal lumen tissue (antiulcer actions), decreases the secretion of fluid into the intestine, absorbs or neutralizes bacterial toxins, inhibits any bacterial activity (antidiarrhea actions), and also increases alkaline secretion to counteract any acid production (antacid action). Thus, this agent controls the frequent voluminous loss of watery stools while relieving intestinal cramping and irritation.

ANTIFLATULENCE AGENTS

Many individuals throughout human history have expressed complaints related to gastrointestinal gas. The most common symptoms of gas are belching, abdominal bloating, abdominal pain, and flatulence. Flatulence is the condition of possessing excessive stomach or intestinal gas. It is a physiological excretory phenomenon caused by the production of gas by bacteria within the colon of the large intestine. The gas is subsequently released from the rectum through the action of abdominal muscles, often resulting in discomfort and social embarrassment for the individual.

Gases are produced by chemical and bacterial or fungal activities in the intestines. The primary components of gas (known as flatus) are odorless gases: including carbon dioxide, hydrogen, methane, and oxygen. The proportions of these gases depend largely on the species of bacteria that reside in the human colon that digest, or ferment, food that has not been absorbed by the gastrointestinal tract through normal digestive processes before reaching the colon. The presence of hydrogen and methane gases contributes to the flammable characteristic of flatus. The characteristic "rotten egg" or ammonia-based odor of flatus may be attributed to the presence of trace gases such as skatole, indole, volatile fatty acids, and sulfur-containing compounds (e.g., hydrogen sulfide ammonia). This is usually the result of putrefaction whereby bacteria have caused proteins to disintegrate.

Compared with foods that contain carbohydrates, foods that contain fats and proteins cause little gas. An estimated 30 to 150 grams of undigested food in the form of carbohydrate may enter the colon of an individual daily. The human body does not easily digest and absorb some types of carbohydrates (e.g., sugars, starches, and fiber) in the small intestine because of a shortage or lack of particular digestive enzymes. Thus, the undigested food passes into the large intestine, where harmless and normal bacteria and fungi break down the food, producing undesirable flatus. During the process of fermentation, yeast fungus (e.g., *Candida*) within the colon may degrade the carbohydrate, producing alcohol and a copious amount of gas. Certain types of foods that may elicit this response include broccoli, cabbage, brussel sprouts, beans/legumes, cauliflower, onions, prunes, corn, wheat, dark beer, red wine, complex carbohydrates such as potatoes and noodles, and sugars such as sorbitol (a sweetener found in sugar-free products), lactose (milk sugar), fructose (fruit sugar), and raffinose (legume, vegetable, and whole grain sugar). While the presence of excess intestinal gas may be attributable to simply eating quickly and swallowing excess air, other physiological conditions that may lead to excessive flatulence include lactose intolerance, constipation, poor dietary insoluble fiber intake, parasites, inflammatory bowel disease, intestinal obstruction (e.g., cancerous tumor), diverticulitis, hypothyroidism, and narcotic/drug use.

While dietary modification may be beneficial for some individuals, over-the-counter oral medications for flatulence treatment are often chosen, which may include simethicone or digestive enzyme supplements. Although activated charcoal or peppermint spirits are also useful in treating flatulence, they are not well tolerated by many individuals. Simethicone, a simple methoxylated siloxane, is a commonly used antiflatulent. This drug acts on the surface of gas bubbles within the intestine, reduces the surface tension, and thereby disrupts (breaks) the bubble film surface, allowing air to escape in the form of belching or passing flatus. This mechanism is accomplished through the action of bridging the liquid film by polydimethylsiloxane droplets, assisted by hydrophobic silica particles also present in this antifoaming, antiflatulence agent.

Antiflatulence digestive enzyme supplements include products containing the enzyme α-galactosidase. This naturally occurring enzyme hydrolyzes the galacto-oligosaccharide sugars (e.g., raffinose, stachyose, and/or verbascose) found in a variety of products ingested by individuals with a high-fiber diet, including grains, cereal, nuts, seeds, soy products, legumes, and vegetables. Thus, unlike simethicone, these types of antiflatulence products actually prevent the production of gas resulting from the ingestion of certain foods before the bacterial fermentation of these foods within the intestine.

ANTIFUNGUS FOOT TREATMENTS

Athlete's foot is a relatively common skin infection of the feet caused by moldlike fungi called dermatophytes. Also known as tinea (the medical word for fungus) pedis (the Latin word for foot), this brief (or potentially long-term) condition is associated with symptoms such as intense itching of feet, cracked, blistered, or peeling areas of the skin (especially between the toes), and redness and scaling on the foot soles. Fungal or tinea infections may also be referred to as ringworm, perhaps because of the scaly, ring-type rash associated with fungal skin infections. The three main types of tinea pedis infections are toe web (interdigital) infection, moccasin-type infection, and vesicular infection. While tow web infection usually involves mild scaling or maceration (severe breaking down) of the skin between the fourth and fifth toes, moccasin-type infection causes scaly thickened skin on the sole and heel of the foot. Vesicular infection involves blistering on the skin of the instep, between the toes, on the sole of the foot, on the top of the foot, or on the heel, causing severe inflammation. Both toe web and vesicular infections may be associated with additional bacterial infection. In some individuals, the fungal infection may also spread to one or more toenails (called onychomycosis), causing the nail to appear unusually thickened and cloudy yellow. Onychomycosis is often associated with moccasin-type tinea pedis infection.

Susceptibility to tinea pedis is increased by poor hygiene, occlusive footwear, prolonged moist skin conditions, and minor skin and nail injuries. Such tinea infections are contagious and can be transmitted through direct contact with infected skin or contact with items such as shoes, stockings, or socks. Also, many cases of tinea pedis may be traced to the use of public recreational or sports facilities and direct skin contact (e.g., walking with bare feet) with warm and damp fungus-contaminated areas around locker room showers, spas, and swimming pools. Because the infection was common among athletes who frequently used such facilities, "athlete's foot" became an increasingly popular term for tinea pedis. Athlete's foot may also spread to other parts of the body of an infected person, notably to the groin (tinea cruris; "jock itch") and underarms, by contaminated bedding or clothing and by scratching the foot infection and then touching elsewhere on the body.

While the top layer of human skin normally hosts a variety of microorganisms, including dermatophytes, advantageous environmental conditions may cause these organisms to multiply rapidly and subsequently cause various types of infections. The dermatophytes that cause tinea pedis thrive in warm, moist environments. The dermatophytes that cause athlete's foot may live on horny nonliving tissues of the hair, nails, and outer skin layers, as the fungus lacks the ability to penetrate the viable tissues of an immunocompetent host. Dermatophytes include three genera (*Epidermophyton*, *Trichophyton*, and *Microsporum*) of fungi that commonly cause skin disease in humans and other animals. Of the three dermatophyte genera, *Epidermophyton* is a filamentous multicellular fungus that includes the pathogenic species *Epidermophyton floccosum*, which specifically causes athlete's foot.

Over-the-counter antifungus (fungistatic or fungicidal) medications typically include active ingredients such as allylamines (e.g., butenafine [benzylamine derivative], terbinafine), azoles (e.g., clotrimazole, miconazole), thiocarbamates (e.g., tolnaftate), hydroxypyridinones (e.g., ciclopirox [$C_{12}H_{17}NO_2$]), or weak acids (e.g., undecylenic acid [$CH_2=CH(CH_2)_8CO_2H$]). These cream, lotion, powder, gel, or spray medications are typically applied topically to the infection site multiple times per day for at least several weeks.

Fungi possess a cell wall in addition to a plasma membrane composed of ergosterol, a key membrane lipid. Both terbinafine and butenafine inhibit ergosterol biosynthesis via the inhibition of squalene epoxidase, an enzyme required for fungal plasma membrane ergosterol synthesis, and subsequently prevent the fungal organism from maintaining a normal intracellular environment. While typically used in concert with a keratolytic agent (e.g., salicylic acid plus benzoic acid) that aids in removing the keratinous layer to allow increased penetration of the antifungus agent, tolnaftate also inhibits squalene epoxidase, resulting in the accumulation of intracellular squalene and decreased ergosterol biosynthesis. Tolnaftate distorts the fungal hyphae and stunts mycelial growth. Conceptually related to allylamines, azoles such as miconazole inhibit the cytochrome P450 microsomal enzyme system and also decrease the conversion of 14-α-methylsterols to ergosterol, thereby inhibiting ergosterol biosynthesis. Clotrimazole, although related to other azoles, acts on fungal cell membranes and interferes with amino acid transport into the organism.

Ciclopirox acts by chelation of polyvalent cations (Fe^{3+} or Al^{3+}), resulting in the inhibition of the metal-dependent enzymes that are responsible for the degradation of damaging peroxides within the fungal cell. A mixture of undecylenic acid and its zinc salt is also used as an active ingredient in athlete's foot remedies. A natural component of human sweat, undecylenic acid has both antifungus and astringent (skin-drying) properties, and the addition of zinc undecylenate liberates undecylenic acid upon contact with areas of moisture. While the fungistatic mechanism of action of this compound is not fully known, possible mechanisms include interference with fatty acid biosynthesis (which can inhibit fungal hyphae [germ tube] formation) and disruption of fungal cell cytoplasm pH by serving as a proton carrier.

COLD SORE TREATMENTS

Herpes labialis (cold sore or fever blister) is a painful and highly contagious infection caused by the herpes simplex type 1 virus (HSV1). The herpesviruses are a large group of double-stranded DNA viruses that cause a variety of diseases in humans and animals. One of the interesting features of some herpesviruses is their ability to remain latent ("lying dormant") in the body for long periods of time after the primary infection, becoming reactivated under conditions such as exposure to heat or cold, hormonal or emotional disturbances, gastrointestinal disturbance, certain foods, stress, fever, infection, common cold, allergies, sunburn (ultraviolet light exposure), female menstruation, pregnancy, or by other unknown stimuli. Latent herpes infections are apparently quite common, with the virus persisting in low numbers within the nervous system via ascending sensory nerve fibers, emerging within epithelial cells in the skin or mucus membranes possibly years later at or near the original site of viral entry. Thus, while the immune system of a healthy individual responds to an initial attack by HSV1 and resolves the symptoms, the virus is able to maintain a small colony of live viruses within the individual, thereby posing a risk of viral reactivation and subsequent additional cold sore formation.

An HSV1 infection (whether primary or reactivated) is usually characterized by the formation of a small group of superficial blistering vesicles (cold sores, fever blisters) appearing on the gums, outside of the mouth and lips, nose (nostrils), cheeks, or fingers (blister stage). The vesicles (blisters) first form and then break open and ooze fluid (weeping stage), leaving a painful reddish ulceration whose surface soon takes on a gray appearance. An exterior yellow crust then develops and eventually sloughs off (crusting stage), revealing new skin underneath (final healing stage). Although the incubation period of HSV1 infection is short (three to five days), untreated cold sores usually remain for approximately seven to ten days, often requiring up to twenty-one days for complete skin healing. Although cold sores generally are not considered seriously dangerous, the infection may be life threatening for individuals whose immune system is depressed by disease or medications. HSV1 may be contracted by children over six months old and may lead to serious illness. In addition, a cold sore infection may cause blindness in both adults and children if the virus spreads to the eye area, and it may lead to the formation of genital lesions if the infection is spread directly or indirectly to the genitals.

Viruses (from the Latin *virus*, referring to poison) are nonliving obligate intracellular parasites composed of protein and nucleic acid (DNA or RNA) that manipulate the host cell to produce and manufacture more viruses. Viral infection occurs by the attachment of virus particles to specific cell receptors within the host cell. After fusion of the host cell plasma membrane with the virus outer envelope, the protein-based viral nucleocapsid (containing the viral DNA) is transported to the host cell nucleus, where components of the viral particle inhibit macromolecular synthesis by the host cell. Herpes viral DNA and new viral nucleocapsid synthesis occurs within the host nucleus, with the acquisition of new viral envelopes via a budding process through the inner membrane of the host nucleus. The mature newly synthesized viral particles are subsequently released through an intracellular membranous organelle system to the outside of the cell and are capable of infecting other host cells.

On initial contact with HSV1, it is estimated that more than 90 percent of humans develop primary herpes. After the primary infection, most humans usually develop specific antibodies. However, despite the presence of these protective antibodies, many adults have repeated cold sore infections. Sores may develop as late as twenty days after the initial exposure to the virus. Herpes viruses are considered extremely contagious. In ancient Rome, an epidemic of cold sores prompted Emperor Tiberius to ban kissing in public ceremonies. It is estimated that more than 100 million episodes of recurrent cold sore outbreaks occur annually within the United States alone.

Viral contact may occur directly through skin-to-skin interaction with cold sores or other herpes lesions (e.g., kissing, hand touching) or through contact with infected razor blades, towels, dishes, utensils, toothbrushes, and so on. Suggestions for the prevention of viral infection include avoiding direct contact with cold sores or herpes lesions in general, washing items that come in contact with the cold sore lesion in hot (preferably boiling) water before reuse, washing hands carefully before touching another person after having touched a cold sore, and avoiding precipitating causes if prone to oral herpes outbreaks.

Symptoms of aches, fatigue, and lack of energy frequently accompany cold sore outbreaks, with fever or flulike symptoms and swollen lymph nodes in the neck often accompanying the first HSV1 attack. Signs and symptoms of cold sores include small, fluid-filled blisters on a raised, red, painful area of the skin near the mouth, or on the fingers, with swollen, sensitive gums of a deep red color. Frequently, pain, itching, sensitivity, or a tingling sensation (tingle stage) at the site (called the prodome) precedes the blistering by one to two days. Diagnosis by a medical professional is made on the basis of the appearance and/or viral culture of the lesion, with additional examination possibly showing enlargement of the neck lymph nodes.

While for most adults and children, cold sores generally clear up without treatment in seven to ten days, pain may be alleviated by home remedies such as avoiding spicy or acidic foods, applying ice or warmth to the blisters, or applying a moistened tea bag to the area, as it has been suggested that tannic acid (a chemical found in tea) may possess antiviral properties. Unfortunately, the HSV1 infection itself cannot be "cured." It is recommended that patients wash blisters gently with soap and water and dry thoroughly to prevent secondary bacterial infections.

Over-the-counter topically applied medical treatments often contain anesthetic-like numbing agents that temporarily inactivate nerve endings in the skin, such as phenol (C_6H_6O), camphor, benzyl alcohol (C_7H_8O), lidocaine, benzocaine, or tetracaine, that assist in drying and alleviating burning, itching, and pain. Medications (e.g., lip balms) that moisturize and soften cold sores to reduce cracking and bleeding include emollients such as petroleum jelly, allantoin, light mineral oil, cetyl alcohol, cocoa butter, glycine, modified lanolin, aloe vera gel, jojoba oil, peppermint oil, meadowfoam seed oil, eucalyptus oil, and beeswax. Anti-itch ingredients (e.g., menthol) can also be included, along with skin-soothing vitamins (e.g., α-tocopherol, tetrahexyldecyl ascorbate) and herbal extracts (e.g., echinacea, green tea extract, grape seed extract). Since excessive exposure to ultraviolet light may lead to cold sore formation, many lip balm medications also contain agents that act as sunscreens (e.g., zinc oxide, which also acts as an antimicrobial agent). In addition, metals such as zinc and amino acids such as lysine have been suggested to exhibit antiviral properties; thus, they are thought to have an inhibitory effect on the ability of the herpesvirus particles to form cold sore lesions.

Over-the-counter topically applied virus-inhibiting medications for cold sores may include products containing docosanol ($C_{22}H_{46}O$) (also called behenyl alcohol, behenic alcohol, 1-docosanol, docosyl alcohol, or *n*-docosanol), an antiviral chemical used to treat symptoms (e.g., pain, discomfort) of lipid-enveloped viral infections (including HSV1). Docosanol inhibits fusion between the plasma membrane and the HSV1 envelope, thereby preventing viral entry into cells and subsequent viral replication. Thus, docosanol may lead to a significant reduction in the duration of symptoms associated with HSV1 infection, including pain, burning, itching, and tingling.

CONTACT LENS CLEANING SOLUTION

In humans, a contact lens lies in the conjunctival sac of the eye. In a closed eye, this sac is a slitlike space between the conjunctiva-covered eyeball and the eyelids. Contact lenses are small polymer "bowls" that float on tears superficial to the corneal eye layer and correct existing visual deficiencies similarly to glasses. While the idea of the contact lens was formulated as early as 1508, it was not until the 1800s that contact lenses became a reality. Hard plastic contact lenses were invented around 1936, and although soft lenses were invented in 1960, they were not available on the commercial market until 1971.

The two large groups of contact lenses are hard lenses (hard-flexible; rigid, gas permeable) and soft lenses. Modern hard lenses are made from rigid plastic polymers that readily allow for the movement of oxygen to the eye. They are used to correct irregular corneas and are durable, gas permeable, and require simpler cleaning and care than soft lenses. Soft contact lenses consist of "floppy" plastic composition, including a hydrophobic backbone of either polyethylene or polypropylene and attached hydrophilic chemical chain groups. Soft lenses rest on the cornea of the eye and take on the shape of the cornea. They contain up to 75 percent water and are very flexible, porous, and comfortable, but they require more expensive and time-consuming care and are less permeable to oxygen than hard contact lenses.

The human cornea and lens are both avascular (lacking blood vessels); thus, they require proteins and oxygen supplied by natural tear fluid and oxygen available from the surrounding air. For these reasons, the wearing cycle of a lens type depends mostly on the lens permeability to oxygen. Lens-wearing schedules include lenses formulated for daily wear (cleaned once a day, removed overnight) and extended/continuous wear (remain on the eye for up to thirty days before they require cleaning). Lens replacement cycles include lenses that can be used for up to two weeks and are then discarded (called disposables) and lenses that last for up to three months (called frequent replacement lenses). Lenses require frequent replacement because of the deposition of lens deposits (including proteins) that may potentially disturb vision, cause deterioration of the lens material, or elicit deleterious allergic and immunological reactions in the eye of the consumer. Removal of these deposits requires nontoxic contact lens solutions formulated for cleaning, disinfection (chemical, thermal, or peroxide based), and enzymatic treatment.

For soft contact lenses, a saline solution consisting of boric acid, sodium borate, sodium (or potassium) chloride, and preservatives (e.g., sorbic acid, edetate disodium, or polyaminopropyl biguanide) is often used as a lens rinse after cleaning or chemical disinfection, as a solution during thermal lens disinfection treatment, or as a lens solution to dissolve enzymatic cleaning tablets responsible for removing protein deposits. The solution is often a sterile, pH-buffered, and isotonic aqueous solution that is compatible with the chemistry of human tear fluid. Disinfectant solutions are sterile ophthalmic solutions that frequently contain microfiltered hydrogen peroxide and sodium chloride, are stabilized with phosphonic acid, and pH buffered with phosphates. Lenses are exposed to the disinfectant solution, along with an added neutralizer and enzymatic treatment, to neutralize the tissue-damaging hydrogen peroxide solution before lens wearing and to enzymatically remove protein deposits. Multipurpose disinfecting/cleaning solutions for soft lenses are designed to eliminate the need for a separate enzymatic or daily cleaner by allowing lenses to remain clean and free of protein deposits when used on a daily basis. They include ingredients such as gentle cleaners (including surfactants) that remove daily protein deposits and dirt (e.g., sodium citrate, phosphate buffer, poloxamer 237, hydroxyalkylphosphonate, poloxamine, sodium lauroyl lactylate), disinfecting antimicrobial agents (e.g., polyquaternium-1, myristamidopropyl dimethylamine, polyaminopropyl biguanide, polyhexamethylene biguanide), lubricants (e.g., hydroxypropyl methylcellulose), and agents that assist in balancing the overall solution for lens-wearing comfort and chemically match natural eye tears (e.g., sodium chloride, potassium chloride, borates [sodium borate], mannitol, edetate disodium, boric acid, sorbitol).

Solutions formulated to clean hard gas-permeable contact lenses are sterile and pH buffered. They are soaking solutions and include lens-hydrating (wetting) agents (e.g., cationic cellulose derivatives) and pre-servatives (e.g., edetate disodium, chlorhexidine gluconate).

CORN AND CALLUS FOOT TREATMENTS

Callus (keratosis) formation is usually an excessive accumulation of dead skin cells that harden and form a thickened hyperkeratosis epidermal lesion of an area of the foot. The appearance of a callus on skin is often considered a defense mechanism to protect the foot against excessive or abnormal pressure and friction. They usually occur on areas experiencing underlying mechanical stress, including the ball of the foot (where meta-tarsal foot bones bear human body weight), the heel, and/or the under-side of the big toe (hallux). They may, however, form over any bony foot prominence. A callus may be asymptomatic (have no pain) or manifest as a burning pain, dull ache, sharp shooting or pinching pain, or a bruising sensation. While calluses may be simple thickenings of the epidermal skin (developing as cutaneous horns, heel fissures, pinch/spin calluses, or tylomas), they may in some cases have a deep-seated central "core," known as a nucleation. This particular type of callus is usually located on the ball of the foot (often corresponding to a prominence of the underly-ing bone) and can be especially painful to pressure. This condition is referred to by medical professionals (e.g., podiatrists) as an intractable plantar keratosis. Some common causes of callus formation are high-heeled shoes, obesity, abnormalities of walking gait, fat feet and high-arched feet, excessively long or displaced metatarsal bones, bony promi-nences, hammer toe deformity, loss (atrophy) of fat pad on the underside of the foot, and short Achilles tendon.

Corns, similar to calluses, are also hyperkeratoses (increased amount of built-up skin) of epidermal foot skin that vary in both location and ap-pearance. They are a thickening and hardening of the dead surface layer of the skin in response to pressure, and they usually form on the toes, where the bone is prominent and presses the skin against the shoe, ground, or other bones. As the corn thickens by the recurrent rub of me-chanical friction, it produces irritation in the underlying skin tissue, which may or may not involve bony structures of the toe, and becomes swollen, reddened, and painful. There also may be a deep-seated nucleation, in which the corn is thickest and most painful where the core presses on an underlying nerve. Common locations where corns form are the top surface of the toe, at the tip of the toe, and between the toes. Although there are five basic types of corns (hard corns, soft corns, seed corns, vascular corns, and fibrous corns), the two most common types are hard and soft corns. Hard corns, known as helloma dura, appear as small concentrated areas of hard skin, usually within a wider area of thickened skin or callus, and can be symptoms of feet or toes not functioning properly. Soft corns, known as helloma molles or kissing corns, tend to be whitish and rubbery in texture and appear between toes, where the skin is moist from sweat or from inadequate drying. Common causes of corns include tight-fitting shoes, deformed or crooked toes, tight socks and stockings, a seam or stitch inside the shoe that rubs against the toe, an excessively loose shoe that leads to foot sliding and abnormal pressure, and prolonged walking on a downward slope.

Home remedy and over-the-counter nonmedicated products for callus and corn treatment usually include changing of footwear to allow for proper fit and extra room in the toe box (toe area), adding cushioning to existing footwear to reduce or remove excess pressure, friction, and irritation between the skin and foot bones or between toes and surrounding footwear, use of a pumice stone or other abrasive material to reduce callus or corn thickness (for calluses in concert with maintaining moist and supple foot skin with regular warm water foot soaking and application of moisturizers), application of nonmedicated pads around the callus or corn to relieve pressure, and/or application of moleskin over areas that tend to callus or lead to corn formation. However, many over-the-counter callus- and/or corn-removing medicated treatments include the active ingredient salicylic acid. The aim of these treatments is to remove the area of hard skin from the foot by applying a corrosive material (in a plaster- or rubber-based medicated cushion vehicle) to the affected area approximately every forty-eight hours as needed for up to fourteen days.

Salicylic acid [2-hydroxybenzoic acid; $C_6H_4(OH)CO_2H$] is a colorless crystalline organic carboxylic acid derived from the bark of the white willow tree and wintergreen leaves (in the form of methyl salicylate) and also prepared synthetically from phenol. The large molecule size permits this beta hydroxy acid to remain on the surface of the skin, allowing it to more effectively penetrate and exfoliate within the callus or corn. Known to have keratolytic actions, salicylic acid (often in over-the-counter concentrations of 40 percent) removes the outer layers of hardened skin from the foot, thereby reducing or removing the acute painful symptoms related to the presence of calluses or corns. Often, surrounding skin is protected from unnecessary corrosion with agents such as petroleum jelly during callus or corn salicylic acid treatment.

Specialized over-the-counter ointment treatments for calluses and feet with dry, cracked skin may include (instead of salicylic acid) an endoprotease in a white petroleum jelly vehicle base. Known to improve the softness and density of the callus, the endoprotease breaks the ten end peptide bonds on the protein chains of callus tissue, leaving the normal skin unaffected (as normal skin does not contain this type of long-chain protein). Once treated, a gentle rubbing action with a wet washcloth is sufficient to gradually exfoliate callus tissue.

COUGH MEDICATIONS

The respiratory tract consists of the airways that conduct air to and from the oxygen and carbon dioxide gas-exchange surfaces within the lungs. The upper respiratory system consists of the nose, nasal cavity, and pharynx (throat), and the lower respiratory system includes the larynx (voice box), trachea (windpipe), bronchi, bronchioles, and alveoli of the lungs. A specialized respiratory mucosal layer, a type of protective epithelial tissue that may additionally contain mucus-secreting cells and cilia, lines the portion of the respiratory system that conducts air to the lungs. This airway conducting system begins at the entrance of the nasal cavity, extends through the pharynx and larynx, and continues along the trachea and bronchi to the bronchioles. The cilia continually propel mucus, loaded with dust particles and other types of debris, toward the pharynx and away from the lungs.

Although individuals can generally cough on purpose (a conscious cerebral brain mechanism), coughing is usually a parasympathetic reflex (an unconscious brainstem mechanism) of the body used to clear the lower respiratory system of aspirated particles (particles inadvertently breathed in), mucus, or excess fluids. This protective reflex operates when receptors located within the respiratory mucosa of the larynx, trachea, or bronchi are irritated by toxic vapors, chemical irritants, or mechanical stimuli. The irritation of the respiratory tract receptors results in a sudden explosive ejection of air from the lungs. This relatively complex visceral reflexive action is coordinated by the coughing center neurons located within the medulla oblongata of the brain. A cough begins with a deep breath in, at which point the opening between the vocal cords at the upper part of the larynx (called the glottis) shuts, trapping the air in the lungs. As the diaphragm and other muscles involved in breathing (e.g., intercostal muscles) press against the lung tissue, the glottis suddenly opens, producing an outflow of air that can reach speeds greater than 100 miles (160 kilometers) per hour. In normal situations, most individuals cough once or twice an hour during the day to clear the respiratory airway of irritants. However, when the level of irritants in the air becomes elevated, or when the respiratory system contains excess mucus or fluids, coughing may become frequent and prolonged, thereby potentially interfering with exercise or sleeping patterns. Excessive coughing may also lead to distress if accompanied by dizziness, chest pain, or breathlessness. The majority of coughs are related to the symptoms of the common cold virus or influenza.

Coughs are generally described as either dry or productive. A productive cough brings up sputum. Sputum is a secretion that is produced in the lungs and the bronchi (tubes that carry air to the lungs) and expectorated (spit out) through the mouth via deep coughing. It typically consists of discharges (e.g., mucus, irritants, etc.) from the respiratory passages combined with saliva. A nonproductive cough (also called a dry or hacking cough) fails to bring up sputum. Over-the-counter oral cough treatments can contain active ingredients such as expectorants (drugs that promote expectoration, the process of coughing up and spitting out). As an orally administered expectorant, guaifenesin [3-(2-methoxyphenoxy)-1,2-propanediol; $C_{10}H_{14}O_4$] stimulates receptors in the gastric (stomach) mucosal tissue lining, which then initiates a reflex increase in respiratory tract fluid secretions. This drug aids in loosening sputum and bronchial secretions by reducing adhesiveness and surface tension. By reducing the viscosity of secretions, guaifenesin increases the efficiency of the cough reflex and ciliary action in removing secretions from the trachea and bronchi, increases sputum volume, and lubricates irritated respiratory tract membranes through increased mucus flow. Thus, bronchial drainage is improved, and dry, nonproductive coughs become more productive and less frequent.

Nonproductive coughs attributable to colds or inhaled irritants, which interfere with rest and sleep, may also be treated with oral products containing antitussives. An antitussive is an agent that prevents or relieves coughing by depressing the cough center in the medulla oblongata or the cough receptors in the pharynx, trachea, or lungs. A centrally acting antitussive is an agent that specifically depresses the cough center located within the medulla oblongata. Dextromethorphan hydrobromide (3-methoxy-17-methyl-9a,13a,14a-morphinan hydrobromide monohydrate; $C_{18}H_{25}NO \cdot HBr \cdot H_2O$) is a synthetically produced non-narcotic centrally acting antitussive available in over-the-counter cough medications. Related to other centrally acting antitussives such as codeine, it is a salt of the methyl ether of the dextrorotatory isomer of levorphanol, a narcotic analgesic (a drug that relieves pain). While dextromethorphan hydrobromide has no analgesic or addictive properties like those of narcotic centrally acting antitussives, this drug suppresses the cough reflex by increasing the threshold for coughing to occur. When ingested at the proper therapeutic dosage, it does not depress respiration or inhibit ciliary action within the respiratory airways.

DECONGESTANTS

Several of the bones located within the skull contain air-filled chambers known as sinuses. These sinuses lighten the various skull bones and are lined by a specialized mucus membrane epithelium, which produces mucus that helps to moisten and clean the air in and around each sinus. Specifically, the nasal complex includes the bones that enclose the nasal cavities and the paranasal sinuses. The paired paranasal sinuses (e.g., maxillary, ethmoidal, frontal, palantine, and sphenoidal) all drain their mucus secretions directly (or indirectly in the case of the palantine) into the nasal (nose) cavity. The mucus within the nasal cavity allows inhaled air to become warmed and humidified and also provides a "sticky" medium for dust or microorganisms to become trapped. The movement of the cilia (cellular membrane extensions located within the nasal cavity epithelial lining) usually passes the mucus backward toward the throat, where it is eventually swallowed into the digestive system. Thus, this drainage system normally protects the fragile respiratory tract from exposure to foreign particulate matter.

Sinus congestion (also referred to as "nasal congestion" or "rhinitis") involves the blockage of one or more of the five pairs of paranasal sinus passageways in the skull. The blockage may result from inflammation and swelling of the nasal tissues, obstruction by one of the small bones of the nose (e.g., in the case of a deviated septum), or the presence of excess mucus (e.g., decreased ciliary activity). While acute (short-term) sinus congestion is frequently caused by the common cold virus (e.g., rhinovirus), chronic (long-term) sinus congestion usually results from exposure to environmental irritants such as tobacco smoke, food allergens, inhaled allergens, or foreign matter within the nasal cavity. Sinus congestion impairs the flow of mucus within the sinuses and further predisposes individuals to secondary bacterial infections that usually cause sinusitis. Sinusitis is an inflammation of the sinuses. The inflammatory response process involves the release of numerous inflammatory chemicals into the extracellular fluid space around cells. The liberated chemicals promote dilation (an increase in diameter) of small blood vessels in the vicinity, increase the permeability of local blood capillaries, and cause exudate (fluid-containing proteins) to seep from the bloodstream into the tissue spaces. The presence of the exudate then causes local edema, or swelling. As the swelling occurs, the drainage passageways narrow, the movement of mucus slows, congestion increases, and the individual experiences headaches and a feeling of pressure within the facial bones.

The sympathetic nervous system is a subdivision of the autonomic nervous system (or visceral motor system), which provides automatic, involuntary regulation of smooth muscle, cardiac muscle, and glandular activity or secretions. Sympathetic nerve fibers reach the nasal mucosa, and when stimulated, sympathetic nerve terminals release norepinephrine, which binds to and activates two types of adrenergic receptors [e.g., α1 and α2; receptors activated by adrenaline(epinephrine)], or substances with similar activity, on the vascular smooth muscle fibers. α1 receptors tend to be concentrated on postcapillary venules, which act as capacitance vessels, and decrease mucosal and blood volume when activated. α2 receptors are concentrated on precapillary arterioles and decrease mucosal capillary blood flow when activated. Thus, when α-adrenergic receptors in the mucosa of the upper respiratory system (e.g., within the paranasal sinuses) are stimulated, there is vasoconstriction (decrease in the diameter of blood vessels) of the mucosal capillaries, with additional shrinkage of swollen mucosal tissue. Drugs that target these adrenergic receptors produce effects similar to those produced by natural stimulation of the sympathetic nervous system. These drugs are referred to as being "sympathomimetic," as they mimic the effects of stimulating postganglionic adrenergic sympathetic nerves.

Nasal decongestants are adrenergic sympathomimetic drugs that allow for vasoconstriction within the parasinus mucosal lining, thereby causing a decrease in swelling, release of exudate, and airway obstruction. Decongestants act to provide relief by stimulating the α-adrenergic receptors of the vascular smooth muscle, constricting the dilated arteriolar network within the nasal mucosa, reducing blood flow in the engorged swollen nasal area, and thereby improving ventilation within the nasal cavity. Both topical (e.g., nasal spray) and oral forms of over-the-counter decongestants are currently available. Topical decongestants act as potent α-adrenergic agonists (stimulants) that constrict blood vessels within the nasal mucosal lining, causing a decrease in swelling and improved mucus drainage. Active ingredients in such topical decongestants include oxymetazoline hydrochloride, xylometazoline hydrochloride, propylhexedrine, and phenylephrine hydrochloride. Oxymetazoline hydrochloride (3-[(4,5-dihydro-1H-imidazol-2-yl)methyl]-6-(1,1-dimethylethyl)-2,4-dimethylphenol hydrochloride; $C_{16}H_{24}N_2O \cdot HCl$) and xylometazoline hydrochloride (2-[[4-(1,1-dimethylethyl)-2,6-dimethylphenyl]methyl]-4,5-dihydro-1H-imidazole hydrochloride; $C_{16}H_{24}N_2 \cdot HCl$) are both sympathomimetic imidazoline derivatives. These two drugs are often the active ingredients of long-acting (twelve-hour) nasal sprays, which are recommended for use no longer than three consecutive days. Chronic use of such products, and thus continuous sympathetic tone, creates a certain level of vasoconstriction and relief from swollen nasal tissues; however, constant sympathetic stimulation triggers the brainstem to produce the nasal cycle, leading to a rebound of nasal congestion and stuffiness. This may then lead the individual to continue using the spray long after the recommended duration, which can lead to drug dependency. It is the prolonged stimulation of α2-adrenergic receptors that is theorized as a mechanism for this rebound effect. Interestingly, oxymetazoline hydrochloride is thought to be such an α2 receptor agonist. Other sympathomimetic imidazoline derivatives, including naphazoline and tetrahydrozoline, also act as potent vasocon-strictors and may be found as active ingredients within both nasal decon-gestant formulations and eyedrop redness-relief treatments. Propylhexedrine (1-cyclohexyl-2-methylaminopropane;$C_{10}H_{21}N$), a sympathomimetic amine, also acts as an α-adrenergic agonist. Structurally related to the central nervous system stimulant amphetamine [a synthetic compound patterned after the chemical structure of naturally occurring epinephrine(adrena-line)], propylhexedrine replaced amphetamine in popular nasal inhalers starting in 1949 as a drug with weaker stimulant properties.

Phenylephrine hydrochloride ((−)-*m*-hydroxy-α-[(methylamino)methyl] benzyl alcohol hydrochloride; $C_9H_{13}NO_2 \cdot HCl$) is an active vasoconstric-tor ingredient in both topical and oral over-the-counter decongestants. Chemically related to epinephrine and ephedrine, this drug is a synthetic sympathomimetic amine that is considered a relatively selective α1-adrenergic agonist. Pseudoephedrine hydrochloride ([*S*-(*R**,*R**)]-α-[1-methylamino)ethyl]benzene methanol hydrochloride; $C_{10}H_{15}NO \cdot HCl$), an active ingredient of oral decongestants, is an α-adrenergic receptor agonist (sympathomimetic) that produces vasoconstriction by stimulat-ing α-receptors within the mucosa of the respiratory tract and also by causing the release of norepinephrine from storage sites. Pseudoephe-drine shrinks swollen mucus membranes, reduces exudate release and nasal congestion, and thereby increases normal nasal airway exchange. Drainage of sinus secretions is increased, and any mucus obstruction within the eustachian tube (the tube connecting the middle ear to the throat) is alleviated. A physiologically active stereoisomer of ephedrine, the vasoconstriction action of pseudoephedrine is similar to that of ephe-drine but produces comparatively less tachycardia (increased heart rate), increased blood pressure, and central nervous system stimulation. After the oral administration of pseudoephedrine, nasal decongestion usually occurs within thirty minutes and persists for four to six hours. Systemic administration of decongestants eliminates possible damage to the nasal mucosa compared with the use of nasal decongestant sprays.

DIURETICS

The term "diuresis" (*dia* = through, *ouresis* = urination) refers to urine excretion. Drugs called diuretics (from the Greek word *diouretikos*, mean-ing promoting urine) cause an overall removal of fluid from the body by targeting the kidneys and increasing the rate of urine flow and sodium excretion from the body. The physiological goal of diuretic administra-tion is to promote the loss of water in the urine and thereby cause a reduction in blood volume, blood pressure, or both within the individual. Diuretics are often used to treat a variety of clinical situations, including hypertension (high blood pressure), congestive heart failure, kidney failure, and edema (abnormal localized accumulation of tissue fluids). Many diuretics available in over-the-counter formulations are considered relatively mild acting and are marketed to alleviate the feelings of bloating, swelling, puffiness, and fullness often associated with monthly premenstrual and menstrual periods in reproductively capable females and temporary water weight gain in both males and females. Common side effects of these agents may include potassium depletion, hypotension (low blood pressure), dehydration, and hyponatremia (sodium depletion).

The basic blood-processing and urine-forming unit of the kidney is the nephron. Each of the two human kidneys contains more than 1 million nephrons, each of which carries out the processes that form urine. In addition, each kidney contains thousands of collecting ducts, each of which collects urine from several nephrons and conveys it to a medial area of the kidney called the renal pelvis. Each nephron consists of a renal corpuscle (containing a tuft of capillaries called the glomerulus and the associated glomerular [Bowman's] capsule) and a renal tubule. Filtration occurs at the renal corpuscle as blood pressure forces fluid and dissolved solutes out of the blood contained in the glomerular capillaries and into the capsular space and then the renal tubule. This filtration process produces an essentially protein-free, plasma-derived fluid solution, known as filtrate, which is the raw material processed by the renal tubules to form urine. The glomerular filtration rate is the total amount of filtrate formed per minute by the kidneys. The renal tubule is approximately three centimeters long and has three named parts. The renal tubule leaves the glomerular capsule as the coiled proximal convoluted tubule, then makes a hairpin loop turn called the loop of Henle, and then finally twists and winds again as the distal convoluted tubule before it empties the fluid product into a collecting duct. Along the renal tubule, water, urea, and various solutes (e.g., ions including sodium, potassium, chloride, bicarbonate, calcium, ammonium, and hydrogen, and molecules such as creatinine, glucose, amino acids, vitamins, and organic acids) may be reabsorbed back into the bloodstream of the body or secreted from the blood of the surrounding peritubular capillaries through the tubule cells, or from the tubule cells themselves, into the filtrate. Thus, urine that is eventually excreted from the body is composed of both filtered and secreted substances.

Diuretics have many different mechanisms of action, but all of them affect transport activity or water reabsorption along the nephron and collecting duct system. Over-the-counter orally administered diuretics can contain chemically active ingredients such as pamabrom or caffeine (3,7-dihydro-1,3,7-trimethyl-1H-purine-2,6-dione; $C_8H_{10}N_4O_2$). In 1955, the diuretic activity of pamabrom, chemically prepared from 2-amino-2-methyl-1-propanol and 8-bromotheophylline, was characterized. Pamabrom (8-bromo-3,7-dihydro-1,3-dimethyl-1H-purine-2,6-dione compound with 2-amino-2-methyl-1-propanol [1:1]; $C_{11}H_{18}BrN_5O_3$) is therapeutically considered a drug combination, meaning a single preparation containing two active ingredients for the purpose of a concurrent administration as a fixed-dose mixture. Chemically, it contains both a propanolamine (amino alcohol) group and a theophylline group. Although caffeine was first synthesized in 1895, the stimulatory actions of caffeine have been known since ancient times. Structurally, caffeine is a methyl xanthine, in the same class of compounds as theophylline. Xanthine itself is a nitrogenous compound (dioxypurine) and is structurally related to uric acid. Theophylline and caffeine are closely related alkaloids that occur in plants widely distributed geographically. For example, tea prepared from the leaves of *Thea sinensis*, a bush native to southern China and now extensively cultivated in other countries, contains caffeine and theophylline. Caffeine is also found as a natural component of cocoa and chocolate, coffee (often extracted from the fruit of *Coffea arabica* and related species), and cola-flavored drinks (contents extracted from the nuts of *Cola acuminata*). Methylated xanthines, theophylline, and caffeine inhibit the 3'5'-cyclic nucleotide phosphodiesterase enzyme that normally degrades cyclic adenosine monophosphate (cAMP). cAMP is important in a wide variety of metabolic responses to cell stimuli, including signal transduction. Thus, the pharmacological actions of these compounds are mediated through a generalized increase in cAMP levels within specific cells of the body to cause various effects, most frequently within organs of the cardiovascular, respiratory, and nervous systems.

Caffeine, and the theophylline chemical group of pamabrom, increases renal (kidney) blood flow and glomerular filtration rate and may also decrease proximal tubular reabsorption of sodium and water, causing a mild diuretic effect. The amino alcohol group of pamabrom may also function as a diuretic, as alcohols act indirectly as diuretics by suppressing the release of antidiuretic hormone from the posterior pituitary gland. After release from the posterior pituitary gland, antidiuretic hormone normally travels through the bloodstream and targets the distal convoluted tubules and collecting ducts of the kidney nephrons to promote water retention, thus decreasing urine output. The diuretic effect of both caffeine and pamabrom is directly dependent on the amount of drug consumed and the duration of intake.

EAR DROPS

The human external (outer) ear includes the external auditory canal (ear canal) and the surrounding auricle or pinna (fleshy cartilaginous shell-shaped projection). The external auditory canal is a short tube (about 2.5 centimeters long by 0.6 centimeters wide) that serves as a passageway between the auricle and the tympanic membrane (eardrum). Near the auricle, the canal consists mostly of elastic cartilage, whereas the remainder of its framework is carved into the temporal bone of the skull. The entire canal is lined with skin bearing hairs, sebaceous (oil) glands, and modified apocrine sweat glands called ceruminous glands that secrete a yellow-brown waxy product called cerumen. Cerumen consists of nearly 50 percent lipids, with additional proteins and carbohydrates. In general, the combination of cerumen, oily secretions from sebaceous glands, squamous epithelial cells from the ear canal lining, and dust and other debris constitutes earwax. The combination of hairs and cerumen naturally assists in slowing the growth of microorganisms (reducing the chances of infection) and in preventing water and damaging debris and foreign objects from entering the ear. Different individuals vary considerably in the amount and consistency of their earwax. In many individuals, the ear is naturally cleansed as the cerumen dries and falls out of the external ear canal. However, some individuals experience excessive cerumen buildup, possibly attributable to abnormal cerumen production. Other predisposing factors leading to an overproduction of cerumen include hairy ear canals, narrow ear canals, and the presence of in-the-ear hearing aids. The excess cerumen may become impacted and cause discomfort, pain, and hearing impairment by the muffling of incoming sounds. The treatment for mildly impacted cerumen is usually periodic ear irrigation, but severe cases can require treatment by trained medical personnel, including ear syringing, physical removal of wax using a blunt instrument (ear curette), and/or suction.

Over-the-counter ear drops used to soften and loosen small to moderate amounts of earwax are all basically oil and peroxide solutions, generally containing active ingredients called ceruminolytics. The drops are placed into the ear canal using a specialized applicator when the head is tilted sideways. The drops remain in the ear by keeping the head tilted or by placing cotton loosely in the ear. A commonly used ceruminolytic includes carbamide peroxide suspended in an organic base (e.g., anhydrous glycerin). Hydrogen peroxide does not directly dissolve earwax, but it may be present for a mechanical effect. "Enzyme"-based ceruminolytic preparations (e.g., containing triethanolamine polypeptide oleate) act as cerumen-lysing surfactants but are seldom chosen by consumers because of their potential to elicit allergic reactions. Other substances often used as "home-remedy" ceruminolytics include a 10 percent aqueous sodium bicarbonate solution (advertised as early as 1860 as an ideal ceruminolytic), glycerin, and olive oil or almond oil warmed to room temperature.

EYE DROPS

In humans, the adult eyeball measures approximately one inch (2.5 centimeters) in diameter, and of its total surface area, only the anterior one-sixth is exposed to the external environment. This exposed area is protected by accessory structures of the eye, including the eyelids, eyelashes, eyebrows, lacrimal (tearing) apparatus, and extrinsic eye muscles. Each eyelid possesses a conjunctiva, which is a thin, transparent, and protective mucus membrane. The conjunctiva lines the inner aspect of the eyelids as the palpebral conjunctiva and folds back (reflects) over the eyelids onto the anterior exposed surface of the eyeball as the bulbar (ocular) conjunctiva. The bulbar conjunctiva covers only the "white" of the eye (not the cornea that covers over the iris and pupil). Because the bulbar conjunctiva is quite thin, blood vessels are clearly visible beneath it. When these blood vessels are dilated and congested from local irritation or infection (conjunctivitis, or inflammation of the conjunctiva), the result is the appearance of reddened and irritated "bloodshot" eyes.

Many consumers purchase over-the-counter eye drop medications for eye symptoms such as redness and/or itchiness or dry eyes. The "redness-reliever" products are generally appropriate for clearing up red or bloodshot eyes that have been exposed to general minor irritants such as smoke, chlorine, and wind. These products usually contain an active redness-reliever ingredient such as ephedrine hydrochloride (α-[1-(methylamino) ethyl]benzyl alcohol hydrochloride; $C_{10}H_{15}NO \cdot HCl$) that occurs naturally in Ma-huang and several other *Ephedra* plant species, naphazoline hydrochloride [2-(1-naphthylmethyl)imidazoline; $C_{14}H_{14}N_2 \cdot HCl$], phenylephrine hydrochloride ((−)-*m*-hydroxy-α-[(methylamino)methyl]benzyl alcohol hydrochloride; $C_9H_{13}NO_2 \cdot HCl$), or tetrahydrozoline hydrochloride [2-(1,2,3,4-tetrahydro-1-naphthyl)-2-imidazoline; $C_{13}H_{16}N_2 \cdot HCl$] ranging in concentrations from 0.25 to 0.012 percent. Their primary mechanism of action is vasoconstriction (decrease in blood vessel diameter), which is accomplished by direct stimulation of specific receptors (α-receptors) on blood vessels of the eye. These drugs act as adrenergic "sympathomimetic" agents, targeting adrenergic receptors within the vasculature of the eye and mimicking the effects of the sympathetic division of the autonomic nervous system. These agents may also decrease aqueous humor flow rate within the eye. Aqueous humor, a liquid found in the anterior cavity of the eye, helps nourish the lens and the cornea. In addition, lubricants (e.g., polyethylene glycol, polyvinyl alcohol, purified water, glycerin) and preservatives (e.g., benzalkonium chloride, edetate disodium, disodium EDTA, boric acid, sodium borate, sodium chloride, sodium hydroxide, hydrochloric acid) can be added to moisturize eye dryness, ensure product purity, and adjust pH. Specialized antihistamine eye drops usually contain an added ingredient such as pheniramine maleate (2-[α-(2-dimethylaminoethyl)benzyl]pyridine maleate; $C_{16}H_{20}N_2 \cdot C_4H_4O_4$), a propylamine-derivative first-generation antihistamine that antagonizes the action of the chemical mediator histamine, and are generally recommended for eyes that are itchy and inflamed as a result of dust, pet dander, pollen, or other allergens.

Natural human tears (or lacrimal secretions) are solutions produced by the lacrimal gland of each eye. Each lacrimal gland lies within the orbit above the lateral end of the eye and releases tears through the excretory lacrimal ducts that empty the tears onto the surface of the conjunctiva of the upper eyelid. From there, the tears pass medially over the anterior surface of the eyeball through the action of blinking. Tears eventually drain away from the eye through a series of cavities and ducts into the nasal cavity. Tears consist of a watery solution containing salts, some mucus, antibodies, and a bactericidal enzyme called lysozyme. Thus, tears cleanse and protect the eye surface as they moisten and lubricate, allowing for clarity of vision. The lacrimal glands become less active with increasing age, so the elderly are more susceptible to eye dryness, infection, and/or irritation. Eye lubricants, or artificial tears, are often used to treat dry eyes by adding moisture and lubrication. Dry eye symptoms include burning, stinging, or a sensation of grittiness caused by a variety of factors, including smoke, fumes, dust, airborne particles, and changes in external temperature and humidity. Over-the-counter eye medications indicated for dry eyes can contain active eye-lubricating ingredients within a saline solution base. These ingredients may include polyethylene glycol, glycerin, hydroxypropylmethylcellulose, carboxymethylcellulose sodium, dextran, polyvinyl alcohol, and povidone. In addition, preservatives that are chemically similar to those found in eye redness-reliever treatments are often included in these formulations. Ophthalmic ointments obtained without prescription may also be used to treat dry eyes. These medications are applied to the inside of the eyelid and tend to treat symptoms for a longer duration than traditional eye drop treatments. However, they tend to blur vision and are mostly used as an overnight treatment. Such eye-lubricating ointments include active ingredients such as white petroleum and mineral oil in a ratio of approximately 80 to 20 percent.

FEMININE HYGIENE TREATMENTS (1)

The term "vulva" (or "pudendum") refers to the external genitalia of the human female. The vulva includes the mons pubis, labia, clitoris, and structures associated with the vestibule. Located anterior to the vaginal and urethral openings, the mons pubis is a rounded elevation of adipose tissue covered by skin and coarse pubic hair that overlies and protects the underlying pubic symphysis. Extending inferiorly and posteriorly from the mons pubis are two sets of elongated labia ("lips"), the labia minora (thin hair-free skin folds that contain oil glands) and the enclosing outer labia majora (hair-covered fatty skin folds that contain oil and apocrine sweat glands). The clitoris is a small, cylindrical mass of erectile tissue and nerves located at the anterior junction of the labia minora. A layer of skin called the prepuce is formed at the point where the labia minora anteriorly unites and covers the body of the clitoris. A central recessed space bounded by the labia minora is called the vestibule. The vestibule contains the external opening of the urethra (exit site of urine) and the vagina (birth canal; the female organ of copulation). Glands located within the vestibule include the lesser vestibular glands, the greater vestibular glands (also called Bartholin's glands), and the paraurethral glands (also called Skene's glands). These glands release mucus into the vestibule, helping to maintain a moist, lubricated vestibular environment. The vestibule is also moistened by secretions originating from the cervical mucus glands located in the inferior region of the uterus. These secretions are metabolized to acidic products by resident bacteria within the vagina. These acidic products are subsequently released through the vagina, and this low-pH environment serves as a natural defense against pathogenic infection.

Pruritus vulvae is a condition characterized by persistent itching of the vagina and frequently the surrounding vulva. Many of the causes of vulvar and vaginal itching are related to irritation of the skin and may elicit symptoms such as genital area pain and burning in addition to itching. Common causes include exposure to chemical irritants (e.g., detergents, fabric softeners, chemical sprays) or medications, vaginal infection in conjunction with antibiotic use, vaginal yeast infection (monilial vaginitis), vaginal discharge caused by infections from *Trichomonas vaginitis* or bacterial vaginosis, diabetes mellitus, vulvovaginitis before puberty, menopause (decreasing estrogen levels), pinworms, lichen sclerosis, or certain vulvar skin conditions. Factors that cause vulvar and/or vaginal itching frequently are attributable to a pH change in the normal acidic vaginal environment. While particular over-the-counter steroid creams or ointments (corticosteroids) may be effective in treating symptoms relating to tissue inflammation, other anti-itch preparations are available to treat other symptoms related to vulvar and/or vaginal itching. Topically applied preparations including creams and ointments often contain analgesics/local anesthetics (e.g., benzocaine [ethyl 4-aminobenzoate]), antiseptics (e.g., resorcinol [formaldehyde-substituted carbomonocycle resin]), skin-soothing and -moisturizing herbal preparations (e.g., *Aloe barbadensis*) and vitamins (e.g., cholecalciferol, retinyl palmitate, tocopheryl acetate), or antihistamines that relieve minor irritation, itching, and soreness.

Other over-the-counter preparations that relieve minor vulvar and vaginal irritation, itching, and soreness include douches. A vaginal douche is the process of rinsing or cleansing the vagina by forcing water or another solution into the vaginal cavity to flush away vaginal discharge or other contents. Containing ingredients such as water, pH-regulating chemicals (e.g., vinegar [acetic acid], citric acid, or baking soda [sodium bicarbonate]), cleansing antimicrobial agents (e.g., povidone-iodine [polyvinyl pyrrolidone complexed with iodine], octoxynol-9), and various fragrances, these preparations are used after menstruation or sexual intercourse or to clear away vaginal secretions and reduce vaginal odors.

FEMININE HYGIENE TREATMENTS (2)

The human female vagina ("sheath") is a tubular, elastic, fibromuscular organ measuring approximately eight to ten centimeters (three to four inches) in length. Lined internally with a mucus membrane, it lies between the bladder and the rectum. It is directed superiorly and posteriorly, extending from the cervix of the uterus to the vestibule, a space surrounded by the female external genitalia. Often called the birth canal, the vagina provides a passageway for delivery of a fetus and for menstrual flow. As the female organ of copulation, the vagina receives the male penis and semen during sexual intercourse. The vaginal mucosal lining contains no glands, but it is lubricated by secretions from the cervical mucus glands. The cervical mucus glands release copious amounts of glycogen (a complex carbohydrate), which is subsequently anaerobically metabolized to organic acids (e.g., lactic acid) by resident vaginal bacteria. Thus, because of the presence of organic acids, the vagina possesses an acidic pH, which allows for a healthy vaginal environment by prohibiting the growth of various pathogens.

An inflammation of the vaginal canal, known as vaginitis, is often the result of vaginal colonization by fungi, bacteria, or parasites. Vaginal discharge, odor, irritation, and burning and/or itching generally characterize this condition. Vaginal yeast infection (also called vulvovaginal candidiasis) is a common cause of vaginal irritation. While frequent symptoms of yeast infection in women are itching, burning, and irritation of the vagina and vulva (external genitalia), painful urination and/or sexual intercourse are common as well. In addition, mostly odorless watery to thick curdy vaginal discharge of whitish-gray coloration (typically described as cottage cheese-like in nature) can be present. Adverse symptoms only appear with yeast overgrowth, as yeasts are typically present in the vagina in small quantities. Because the acidic environment of the vagina prohibits yeast overgrowth, biological factors that lead to acidic imbalance within the vagina typically allow for the formation of vaginal infection. Such factors include menstruation, pregnancy, diabetes mellitus, compromised immune function, specific antibiotics, birth control pills, and steroids. Moisture, perspiration, and irritation caused by the use of particular over-the-counter feminine hygiene products, or by wearing tight, poorly ventilated clothing and undergarments, may also promote vaginal yeast colonization.

A yeastlike fungus called *Candida albicans* causes vaginal yeast infections. As a normal inhabitant of humans that normally causes no ill effect, this fungus is commonly found in the mouth, intestinal tract, and vagina. Although *C. albicans* is the pathogen often identified in women with vulvovaginal candidiasis, other possible pathogens include *Candida tropicalis* and *Candida glabrata.* As unicellular fungi inhabiting a variety of natural habitats, yeasts typically multiply as single cells that divide by budding or direct division, or they may grow as simple irregular filaments (mycelium). In sexual reproduction, most yeasts form asci, which contain up to eight haploid ascospores. These ascospores may fuse with adjoining nuclei and multiply through vegetative division; in some yeasts, they fuse with other ascospores.

Various over-the-counter antifungus medications, in the form of creams, tablets, or suppositories, are available to treat vaginal yeast infections. These agents are typically inserted into the vagina with a plunger-type applicator and massaged into the vulva for one to seven days depending on the formulation. Such products usually contain one of three active antifungus ingredients: butoconazole nitrate, miconazole nitrate, or tioconazole. Approved as over-the-counter drugs by the Food and Drug Administration in the 1980s and 1990s, these drugs are characterized as within the same antifungus chemical family (imidazoles) and function similarly in destroying the *Candida* organism. As azole derivatives, imidazoles inhibit cytochrome P450 and lanosterol 14α-demethylase enzyme activity within fungi. With respect to antifungus activity, this action then prevents hydroxylation of the 14α-methyl group of lanosterol, thus decreasing the conversion of 14α-methylsterols to ergosterol, an important membrane lipid component of fungi. Failure of ergosterol synthesis, and additional accumulation of 14α-methylsterols, causes altered membrane permeability, leading to loss of the ability to maintain a normal cellular environment and subsequent growth arrest. In addition, the azole antifungals inhibit morphogenetic transformation of yeast blastospores into the invasive mycelium form in *C. albicans.*

Clotrimazole, an imidazole antifungus drug, may also be included as an active ingredient in over-the-counter treatments for vaginal candidiasis. This agent is related to other azoles but elicits a different mechanism of action in destroying invasive fungi, such that it acts on fungal cell membranes and interferes with essential amino acid transport into the organism.

HEMORRHOID TREATMENTS

The term hemorrhoids, often called piles, refers to a condition in which clusters of veins just under the membrane that lines the lowest portion of the anus or lower rectum become swollen and inflamed. They are thought to represent the engorgement or enlargement of the normal fibrovascular cushions lining the anal canal. Clusters of vascular tissue (e.g., arterioles, venules, arteriolar-venular connections), smooth muscle, and connective tissue lined by normal epithelium of the anal canal constitute the hemorrhoids normally present in humans from the time in utero throughout normal adult life. Pathologic hemorrhoids are thought to result from chronic straining secondary to constipation, prolonged attempts at defecation, or occasional diarrhea. Hemorrhoids have three main cushions, situated in the left lateral, right posterior, and tight anterior areas of the anal canal, with minor tufts located between the cushions. As an individual strains rectally repeatedly over time, the fibrovascular cushions eventually lose their attachment to the underlying rectal wall, leading to the prolapse of hemorrhoidal tissue through the anal canal. As the hemorrhoids engorge with blood, the overlying mucosa becomes friable and the vasculature increases, leading to rectal bleeding in the form of bright red blood.

Hemorrhoids have plagued humans throughout history, sometimes associated with the assumption of an upright bipedal posture. Beginning in Medieval times, hemorrhoids were known as Saint Fiacre's curse. This legend is based on the story of Saint Fiacre, the patron saint of gardeners. Having developed a terrible case of prolapsed hemorrhoids after cultivating a large patch of farmland in one day, Saint Fiacre was miraculously cured after sitting on a stone and praying for a resolution of his problems.

Hemorrhoids are very common in both men and women and may result from any increase in pressure in the veins of the lower rectum; such sources of pressure include constipation and the accompanying extra straining to move stool, sitting or standing for excessive lengths of time with associated increased resting anal pressure, obesity, heavy lifting, portal hypertension, and pregnancy and childbirth. Other contributing factors may include aging, chronic diarrhea, anal intercourse, and a genetic predisposition to develop hemorrhoids. While the most common symptom of hemorrhoids is rectal bleeding, such bleeding may be associated with many other anorectal problems, including anal fissures (tears in the anus lining), fistulae (abnormal channels that develop between the anal canal and the skin around the opening of the anus), abscesses, pruritus ani (anal irritation and itching), proctitis (an inflammation of the inner lining of the rectum), colon or rectal growths (polyps), colorectal cancer, viral and bacterial skin infections, rectal prolapse (a portion of the rectum protrudes through the anus), or diverticular disease (small sacs or pouches [diverticula] that form from the lining of the large intestine).

Hemorrhoids generally cause symptoms when enlarged, inflamed, thrombosed, or prolapsed. Hemorrhoidal symptoms are divided into internal and external sources of anatomical location. Those originating above the dentate (pectinate) line (higher up in the rectum) are termed internal hemorrhoids, and those originating below the dentate line (arising at the entrance to the anal opening) are termed external hemorrhoids. Because internal hemorrhoids lack pain-sensitive somatic sensory nerve fibers, they usually do not cause cutaneous pain. However, they are associated with rectal bleeding and a feeling of fullness in the rectum after a bowel movement, often becoming more severe with straining and eventually bulging (prolapsing) outside the anal opening to cause a constant dull ache and irritation with itching and bleeding. Prolapsed internal hemorrhoids can cause perianal pain by producing a spasm of the sphincter complex and can deposit mucus onto the perianal tissue, causing localized dermatitis. A grading system is used to describe the severity of internal hemorrhoids: grade 1 (mild distention), grade 2 (prolapse with bowel movement, with spontaneous reduction back into the rectum), grade 3 (prolapse with bowel movement, with manual reduction back into the rectum required), and grade 4 (prolapse without the ability of reduction, with additional prolapse of the rectal mucosa [inner lining]).

Externally prolapsed hemorrhoids tend to be associated with symptoms such as severe pain (caused by their innervations by cutaneous nerves that supply the perianal area, such as the pudendal nerve and sacral plexus), inflammation (swelling), and irritation with itching and bleeding. Often, blood will pool within an external hemorrhoidal vein and form a clot (thrombus), resulting in acute pain from the rapid distension of innervated skin by the clot and surrounding edema (fluid buildup). The pain may last seven to fourteen days and may vacate from resolution of the thrombosis. However, remnants of the thrombosis may persist as excess skin or skin tags and occasionally leads to erosion of the underlying skin and additional bleeding. The skin tags may then cause patient difficulty with maintaining adequate rectal hygiene by mechanically interfering with cleansing of the perianal skin area after a bowel movement. Excessive rubbing or cleaning around the anus may remove healthy and protective mucus, leading to irritation with bleeding and/or itching, producing a vicious cycle of symptoms.

In many cases, hemorrhoids may be prevented by self-care and lifestyle changes, such as eating high-fiber foods (e.g., fruits, vegetables, and grains), drinking plenty of liquids daily, ingesting a bulk stool softener or fiber supplement (e.g., psyllium or methylcellulose), exercising, avoiding long periods of standing or sitting, avoiding lower rectal straining during defecation, and voiding stools in timing with the natural accompanying urge to have a bowel movement.

While surgical intervention may be indicated in some cases of hemorrhoids, temporary relief of the mild pain, swelling, and inflammation symptoms of most hemorrhoidal flare-ups may be accomplished with the use of warm water bath soaking of the affected area several times daily and application of a variety of over-the-counter topical hemorrhoidal cream, ointment, foam, spray, or suppository medications multiple times daily for a limited duration. Such medications can contain active ingredients that provide a physical barrier with additional lubricant qualities (e.g., petrolatum, mineral oil, shark liver oil, cocoa butter, zinc oxide) with additional skin-protecting lubricants (e.g., shea butter, *Aloe vera* gel, simethicone, beeswax, lanolin, glycerin). Phenylephrine hydrochloride is often included, acting as a local vasoconstrictor, thus reducing the symptoms of swelling and itching. The use of topical astringents (e.g., witch hazel) can provide a soothing effect, along with skin-soothing vitamins (e.g., tocopherol acetate) and herbal extracts (e.g., green tea extract, grape seed extract). Topical corticosteroids (e.g., hydrocortisone acetate) may be included to reduce inflammation, itching, and swelling. Bismuth can be combined with such corticosteroids and provides a protective barrier to the irritated area and prevents cutaneous water loss. In addition, many products contain local anesthetics such as benzocaine and pramoxine hydrochloride to relieve acute pain.

INSOMNIA TREATMENTS

Insomnia typically refers to abnormal wakefulness, shortened sleeping periods, difficulty in getting to sleep, or early awakening with an inability to return to sleep. This condition may be a reflection of normal age-related changes (e.g., decline in melatonin hormone production by the pineal gland), illness or physical discomfort, stimulants such as caffeine or drugs, or the effects of jet lag in the case of frequent travelers. However, a common cause of insomnia tends to be psychological disturbance (e.g., stress, depression, anxiety). Mild insomnia may be relieved by a soothing activity such as reading or listening to relaxing music. Chronic (long-term) or severe insomnia usually requires treatment of the underlying physical or psychological disorder, along with the use of sedatives and hypnotic drugs if the sleeplessness impairs the ability to obtain the quantity and quality of sleep needed for the person to both function adequately during the daytime and maintain a sense of well-being. Interestingly, sleep requirements in healthy individuals tend to vary from four to nine hours a day; thus, it is difficult to determine the exact amount of sleep needed for normal functionality by every person.

Many over-the-counter oral sedative-hypnotic drugs targeted for the treatment of mild to moderate insomnia include the active ingredient diphenhydramine hydrochloride. This compound is an H_1 antagonist of the ethanolamine class and acts as an antihistamine by competing with free histamine for binding at H_1 receptor sites. Other members of the ethanolamine class include compounds such as carbinoxamine, clemastine, doxylamine, and phenyltoloxamine. Ethanolamine H_1 antagonists such as diphenhydramine HCl have significant anticholinergic and antimuscarinic activity and produce marked drowsiness and/or sedation in most individuals via depression of the CNS. The altered level of consciousness manifested within the CNS, as caused by diphenhydramine HCl, results from central cortical and subcortical muscarinic receptor antagonism. Diphenhydramine HCl causes both a decrease in sleep latency and an increase in sleep duration (slow-wave sleep). This drug also decreases the rate of rapid eye movement sleep during slow wave (deep) sleep. The degree of CNS manifestation (i.e., sedation) is related to the drug's ability to cross the blood-brain barrier, which normally protects the brain from exposure to particular substances traveling within the bloodstream.

Anticholinergic agents diminish the effects of the neurotransmitter acetylcholine, a substance produced by the body that is responsible for certain nervous system activities. Diphenhydramine HCl provides a competitive reversible blockade of muscarinic receptors, thereby inhibiting cholinergic neurotransmission via acetylcholine at muscarinic receptor sites. The term "muscarinic" indicates that the receptor can be stimulated by muscarine, a toxin produced by some poisonous mushrooms (e.g., *Amanita muscaria*). Muscarinic receptors are located at cholinergic neuroeffector junctions in the parasympathetic and sympathetic divisions of the autonomic nervous system, which is involved with the involuntary control of physiological body processes. The cholinergic neuroeffector junctions, which are types of chemical synapses, involve the release of the neurotransmitter acetylcholine from presynaptic neurons to postsynaptic effector cells to allow for communication between the nervous system and other parts of the body. Muscarinic receptors present at neuroeffector junctions may produce either an excitation or an inhibition response, depending on the nature of the enzymes activated when acetylcholine normally binds to the receptor.

The maximum sedative effect of diphenhydramine HCl usually occurs between one and three hours after oral administration, with the duration of action ranging from four to six hours. Thus, because of its ability to induce drowsiness, diphenhydramine HCl is promoted as an over-the-counter hypnotic to be used as a short-term sleep aid. Prolonged use of diphenhydramine HCl can result in the development of tolerance, which may lead to reduced sedative effects. Therefore, chronic use of such medications is not recommended, as long-term continuation of the recommended drug dose may not alleviate insomnia. In addition, increased blood serum concentrations of diphenhydramine HCl, present as a result of ingestion beyond the recommended dose, can result in general systemic toxicity and overall bodily harm.

LAXATIVES

A laxative is a synthetic or natural drug (or other substance) used to stimulate the peristaltic action of the large intestines in eliminating fecal waste from the body. The term "laxative" usually refers to mild-acting substances, whereas "cathartic" agents are substances of increasingly drastic action. Laxatives are readily available as over-the-counter drug products and as dietary supplements often used to treat constipation. Constipation refers to the infrequent, difficult, or consistently incomplete passage of small hard feces, usually fewer than three times a week. Constipation may be caused by numerous factors, including lack of adequate fiber or fluid in the diet, prolonged physical inactivity, types of medications (e.g., opioid analgesics, barbiturates, antidepressants, antipsychotics, antihistamines, iron, diuretics, anticonvulsants, angiotensin-converting enzyme inhibitors, calcium channel blockers, and calcium- or aluminum-based antacids), metabolic or endocrine disorders (e.g., diabetes mellitus, hypothyroidism, kidney failure), neurological disorders (e.g., multiple sclerosis, lupus, stroke, Parkinson's disease, brain tumor), large intestinal disorders (e.g., hernias, tumors, diverticulitis, irritable bowel syndrome), chronically avoiding the urge to empty bowels, or abuse of laxatives. Symptoms include anorexia, dull headache, low back pain, abdominal distension, and lower abdominal distress.

Ancient Egyptian laxative formulations were known to include different types of salt recipes in liquid, suppository, or ointment form. These included a laxative suppository consisting of honey, vegetable seeds, and ocean salt as well as one of incense, vegetable seeds, fat, oil, and ocean salt. Ancient Arab medicinal laxatives included both senna leaves and pods (fruits). Modern over-the-counter constipation treatments include different types of laxatives. Bulk-forming laxatives allow for the absorption of water from the intestinal lumen, creating a larger stool mass that facilitates evacuation. Once expanded, the stool mass triggers the natural large intestine contraction cycle that leads to expulsion of the feces. They are always administered along with fluids (e.g., water, fruit juice, etc.) and include active fibrous hydrophilic ingredients such as bran, oat, wheat, psyllium seed husks (from *Plantago ovata* and *Plantago major*), agar-agar, calcium polycarbophil, flax seed, apple pectin, or various cellulose substances (e.g., methylcellulose).

Hyperosmotic laxatives exert a local irritant effect and an osmotic effect such that water is drawn into the rectum, which results in abdominal distension and stimulates peristaltic bowel movement. They include active ingredients such as glycerin, lactulose, sorbitol solutions, or polyethylene glycol. Saline laxatives are relatively nonabsorbable cations and anions that draw water into the large intestine, thereby causing an increase in intracolonic pressure and subsequently an increase in intestinal peristaltic motility. These laxatives include active ingredients such as magnesium hydroxide (milk of magnesia), magnesium citrate and magnesium sulfate (Epsom salts), or sodium phosphate.

Emollient laxatives (stool softeners) allow for water to interact more effectively with stool solids, resulting in a softening of stool consistency. Thus, perirectal strain is avoided during evacuation. They add oily material (e.g., mineral oil), increase water retention in the intestine, or aid the mixing of water and oil components (e.g., with assistance from docusates [surfactants]) in the fecal material. Lubricant laxatives act similarly to emollient laxatives in that lubricants (e.g., mineral oil) soften fecal contents by coating the stool and preventing colonic absorption of fecal fluids, including water. Lubricants ease the passage of wastes and counteract excessive drying of the intestinal contents.

Stimulant laxatives directly stimulate colonic smooth muscle to promote intestinal peristalsis and the secretion of water into the bowel lumen. These laxatives include ingredients such as castor oil, casanthranol, bisacodyl, or phenolphthalein (also an acid-base indicator). Herbal stimulant laxatives include senna (both leaves and fruits of *Cassia senna*), cascara (*Casara sagrada*), buckthorn (also known as frangula), rhubarb root, and dried aloe juice. All of these naturally occurring herbal stimulant laxatives (along with danthron, another previously marketed stimulant laxative) contain one basic category of chemical compounds, known as anthraquinones. These include both the simple anthrones and the bianthrones (composed of two anthrones linked together). The bianthrones, especially sennosides (as found in rhubarb and senna), are quite active as laxatives. Frequent or regular (chronic) use of stimulant laxatives should be avoided, as serious disruption of natural digestive processes may occur (e.g., malabsorption of nutrients and vitamins, electrolyte and fluid deficiencies, and lack of natural colonic muscle contraction [called "cathartic colon"]).

MAGNETIC RESONANCE IMAGING

Magnetic resonance imaging (MRI) is a medical technique used to obtain images of body tissue. Physicians use the procedure to examine a wide range of medical problems, including diagnosing diseases (multiple sclerosis), identifying the presence of tumors, identifying infections in the body, visualizing torn ligaments, diagnosing tendonitis, and even imaging the flow of blood in virtually any part of the body. The advantage of MRI is that it allows for the identification of abnormal tissue, caused by injury or disease, without invasive surgery or exposure to radiation (x-ray diagnostic test).

An MRI instrument basically consists of three major components: a large cylindrical magnet, devices to transmit and receive radio waves, and an imaging device (for computer analysis). Although newer models are decreasing in size, the MRI instrument itself is a large machine, typically having dimensions of approximately seven feet high by seven feet wide by ten feet long with a large tube (the Bore) running through the magnet. The magnet is very strong, usually in the range of 0.5 to 2.0 Tesla (the international unit for magnetic flux named for the inventor Nikola Tesla). One Tesla is equal to 10,000 G (Gauss). The earth's magnetic field is 0.5 to 1 G, so the magnet in an MRI machine can be 20,000 times the strength of the earth's magnetic field. The main magnet immerses the patient in a stable intense magnetic field. This cause the atoms of the patient's body to line up in the direction of the magnetic field. MRI specifically targets the hydrogen atoms because of their ideal characteristics. If a patient is lying on his or her back in the magnetic field that runs down the center of the tube of the machine, the hydrogen atoms will line up in the direction of either the feet or the head. Although most hydrogen atom alignments cancel each other out, the large number of hydrogen atoms in the body allows for the analysis and development of detailed images. The next step in the process is to apply a radio frequency pulse (a low-energy pulse that is specific to hydrogen). This pulse is directed to the area to be analyzed. The atoms absorb the energy from the pulse if it is at the same frequency as the radio wave. This causes the hydrogen atoms to precess, or spin, in a different direction. This is the "resonance" step of the MRI process. The exact frequency of the resonance is called the Larmour frequency and is calculated based of the type of tissue being examined and the main MRI magnet. Thus, the physician can tune the machine to examine specific parts of the body. In addition, the MRI system has another type of magnet called a gradient magnet (three, typically). These magnets are very low in strength compared with the main magnet. The gradient magnets are arranged so that when they are turned on and off rapidly they alter the main magnet field at a specific location. In essence, this process allows for exact focusing on a specific point on the body. MRI is capable of examining a precise spot within the body as thin as a few millimeters.

The "imaging" of MRI occurs when the radio frequency pulse is turned off. The hydrogen atoms begin to return to their natural alignment within the field and release the excess stored energy acquired from the radio pulse. This energy is given off as a weak radio signal, which is captured by the coils and sent to the computer system for mathematical analysis (Fourier transform). The strength and length of the energy signal captured depend on various properties of the body tissue; thus, a detailed diagnostic image can be produced.

MOTION SICKNESS TREATMENTS

Motion sickness is a very common disturbance caused by repeated exposure to such motions as the swell of the sea, the movement of a car, or the turbulence of a plane in flight. Although many ancient seafaring nations were familiar with this malady over thousands of years, motion sickness has become increasingly prevalent with the many forms of vehicular travel currently available. Names including seasickness, airsickness, carsickness, train sickness, amusement park ride sickness, flight simulator sickness, and space motion sickness provide an indication of the many causes of this ailment. Characteristically, motion sickness is usually "experienced" by initially sensing gastric discomfort, followed by increased salivation, eructation (belching), and a feeling of general bodily warmth. When continually exposed to a motion sickness-triggering stimulus, normal digestion is decreased and the symptoms of nausea, pallor, and sweating are all increased. Eventually, vomiting or retching occurs. Interestingly, the word "nausea" is derived from the Greek word for ship (*naus*). Another distinct syndrome of motion sickness that lacks gastrointestinal complaints usually includes drowsiness, headache, generalized discomfort (malaise), and various changes in mental perspective. Sufferers have been known to experience a psychological shift in attitude from one of happy excitement to apathy, depression, and nearly suicidal despair in a very short time.

The sensation of balance, and thus the ability to maintain equilibrium, is regulated by a complex interaction of bodily systems that includes the inner ears, the eyes, skin pressure receptors, muscle and joint sensory receptors, and the CNS (brain and spinal cord). The cause of motion sickness is generally considered to be a mismatch of visual and inner ear sensations. The eyes observe where the body is in space (e.g., upside down, right side up, etc.) and also directions of motion. Within the inner ear, these sensations pertain specifically to the vestibular complex, including the vestibule (which includes a pair of membranous sacs containing receptors for the sensation of gravity and linear acceleration) and the semicircular canals (which contain ducts with receptors that are stimulated by rotation of the head and detect angular acceleration). The CNS receives visual and vestibular sensory information and "compares" the inputs with the individual's expectations of motion derived from previous experiences. Motion sickness tends to manifest when central processing stations, such as within the brainstem (including the midbrain, pons, and medulla oblongata), receive conflicting sensory inputs from the vestibular and visual systems (e.g., the vestibular complex detects motion while the passenger is flying in a plane experiencing air turbulence, but the eyes do not detect motion at all as only the inside of the plane is visualized). Motion sickness occurs most commonly with acceleration (forward and backward) in a direction perpendicular to the longitudinal axis of the body (especially when head movements occur away from the direction of motion, as when turning the head backward to face the backseat of a forward-moving car) and with slow-moving vertical (up and down) oscillatory motion called heave (e.g., on camelback or onboard ships). Motion sickness symptoms tend to subside after thirty-six to seventy-two hours of continuous exposure but may return on exposure to the "preexposure" environment until readaptation takes place (this is called *mal de debarquement* syndrome, or arrival sickness).

Nausea and vomiting (also called emesis; from the Greek word *emetikos* meaning inclined to vomit) are the most common complaints of motion sickness and are mediated by CNS chemical messengers called neurotransmitters. In response to visual and vestibular input, increased levels of dopamine stimulate the medulla oblongata's chemoreceptive trigger zone, which in turn stimulates the vomiting center (VC) located within the reticular formation of the brainstem. The chemoreceptive trigger zone is extremely sensitive to the actions of drugs and chemical toxins, mainly because it is not protected by the blood-brain barrier. The VC is also directly stimulated by inputs from the vestibular complex via the vestibular nuclei (a grouping of nerve cell bodies) positioned between the pons and the medulla oblongata and by increased levels of the neurotransmitter acetylcholine. The VC and vestibular nuclei contain muscarinic cholinergic receptors, while the vestibular nuclei also contain histamine H1 receptors (H_1). The name "muscarinic" indicates that the receptor can be stimulated by muscarine, a toxin produced by some poisonous mushrooms (e.g., *Amanita muscaria*).

Most over-the-counter products that are marketed for the prevention or amelioration of motion sickness contain active antiemetic (drugs that prevent nausea and/or vomiting) ingredients, including anticholinergics or antihistamines. When used specifically to prevent motion sickness, these antiemetic medicines generally are most effective if administered well before the motion activity takes place. However, the precise action of these medications in preventing motion sickness is unclear. Histamine receptor 1 antagonist drugs typically act as antihistamines by reversibly competing with free histamine for binding at H_1 receptor sites. In addition, these drugs may also act as anticholinergics/antimuscarinics by competitively antagonizing the binding of acetylcholine at muscarinic receptor sites on postsynaptic neurons. Anticholinergics inhibit the effect of acetylcholine, which normally controls the contraction of skeletal muscles and also plays an important role in the chemistry of the brain and peripheral nervous system. Anticholinergics block acetylcholine stimulation of motility and secretions throughout the entire gastrointestinal tract. In addition, H_1 antagonists block neural pathways originating in the vestibular complex and antagonize muscarinic cholinergic receptors.

Examples of antihistamines used to prevent and treat the nausea, vomiting, and dizziness associated with motion sickness include meclizine HCl, cyclizine HCl, diphenhydramine HCl, and dimenhydrinate. Meclizine HCl (1-[(4-chlorophenyl)phenylmethyl]-4-[(3-methylphenyl)methyl]piperazine dihydrochloride monohydrate; $C_{25}H_{27}ClN_2$ X $2(H-Cl)$x H_2O) and cyclizine HCl (1-diphenylmethyl-4-methylpiperazine hydrochloride; $C_{18}H_{22}N_2 \cdot ClH$) are both piperazine-derivative antihistamines. The antiemetic effect of meclizine HCl is thought to be mediated through the chemoreceptive trigger zone of the medulla oblongata and by the general blocking of muscarinic receptors within the CNS. Cyclizine HCl is believed to block the H_1 and muscarinic receptors associated with the VC, reducing the sensitivity of, and the activity along, the pathway that involves the transmission of nerve impulses from the vestibular complex of the inner ear to the VC. Cyclizine HCl also increases lower esophageal sphincter tone and relaxes the smooth muscles in the stomach directly to prevent emesis. The VC can also receive many excitatory inputs from nerve endings of vagus nerve (cranial nerve X) sensory fibers within the gastrointestinal tract.

Ethanolamine-derivative H_1 antagonist antihistamines used to treat motion sickness include diphenhydramine HCl [2-(diphenylmethoxy)-N,N-dimethylethylamine hydrochloride; $C_{17}H_{21}NO \cdot HCl$] and dimenhydrinate [8-chloro-3,7-dihydro-1,3-dimethyl-1H-purine-2,6-dione compound with 2-(diphenylmethoxy)-N,N-dimethylethanamine (1:1); $C_{24}H_{28}ClN_5O_3$]. Diphenhydramine HCl provides a competitive reversible blockade of muscarinic receptors, thereby inhibiting cholinergic neurotransmission via acetylcholine at muscarinic receptor sites and acting as an antiemetic. It stabilizes the motion-sensitive vestibular complex balance center in the inner ear, thus stopping or preventing the sensation of imbalance caused by effects on the CNS. Dimenhydrinate, a combination of diphenhydramine and chlorotheophylline, also possesses anticholinergic activity and has been shown to depress inner ear vestibular function. Although the exact antiemetic mechanism is unknown, dimenhydrinate, as an H_1 antagonist antihistamine, has been shown to be effective in treating motion sickness by acting as a central antagonist of acetylcholine (anticholinergic). This drug blocks excitatory impulses originating in the vestibular complex of the inner ear at cholinergic synapses in the region of the vestibular nuclei of the brainstem.

Currently, the most popular anticholinergic agent used to treat motion sickness is the centrally acting antimuscarinic/anticholinergic drug called scopolamine hydrobromide (or hyoscine hydrobromide; $C_{17}H_{21}NO_4 \cdot HBr$). An atropine derivative, it is an alkaloid drug obtained from plants of the nightshade family (*Solanaceae*), chiefly from black henbane (*Hyoscyamus niger*). This drug is delivered via a cutaneous patch consisting of a drug reservoir that contains scopolamine, mineral oil, and polyisobutylene (elastomer, or synthetic rubber; C_4H_8) sandwiched between polyester film and an adhesive layer. This dime-sized patch is applied to an area of intact, dry, and hairless skin behind the ear and delivers a slowly absorbed continuous dose of scopolamine into the systemic blood circulation for three days. Structurally similar to acetylcholine, scopolamine prevents motion-induced nausea by interfering with the transmission of nerve impulses by acetylcholine at muscarinic receptors, thereby inhibiting inner ear vestibular nerve stimulation and decreasing nerve transmission and communication to the vestibular cerebellar pathway that terminates in the CNS. This action results in inhibition of the vomiting reflex. Since scopolamine crosses the blood-brain barrier, it may also have a direct action on the VC muscarinic receptors within the brainstem. It also reduces spasms of the digestive system, bladder, and urethra that often accompany the feelings of motion sickness and nausea.

PREMENSTRUAL SYNDROME DRUGS

During their reproductive years, nonpregnant human females usually experience a cyclical sequence of changes in the ovaries and uterus. Each cycle has a duration of approximately twenty-eight days (one "month"; often between twenty-one and thirty-five days in healthy females of reproductive age) and is controlled principally by various protein-based hormones, including gonadotrophin-releasing hormone, follicle-stimulating hormone, luteinizing hormone, and oxytocin, and steroid-based hormones, including estrogen and progesterone. Hormones in general are chemical messengers released by cells into the extracellular fluids (e.g., blood) that regulate the physiological function of other cells within the body. The general term "female reproductive cycle" encompasses the unique events that occur within the ovarian and uterine cycles, the hormonal changes that regulate these cycles, and the additional cyclical changes in the breasts and cervix. While the ovarian cycle specifically refers to a series of events associated with oogenesis and the maturation of the ovum (egg) in the ovaries, the uterine cycle (also called the menstrual cycle) refers to a series of cyclical changes that occur within the inner tissue lining (endometrium) of the uterus. Each "month," the endometrium is prepared for the potential arrival and implantation of a fertilized ovum that will develop in the uterus during a nine-month gestation (pregnancy) period until birth. If fertilization of the ovum by a sperm cell does not occur, a portion of the endometrium tissue layer (called the "stratum functionalis") is shed from the female body through the cervix and vagina. Overall, the endometrial changes of the uterine cycle are coordinated with the phases of the ovarian cycle during each twenty-eight-day "monthly" period through changing levels of hormones in the bloodstream.

The events of the three-stage uterine cycle are menses (menstrual flow phase; days one to five), the proliferative phase (days six to fourteen), and the secretory phase (days fifteen to twenty-eight). The menses phase is marked by the degeneration and detachment of the thick functional tissue zone (stratum functionalis) from the endometrium of the inner uterine wall primarily as a result of decreases in progesterone hormone levels in the bloodstream. The deterioration occurs in patches and is caused by the constriction of arteries, which reduces the blood flow to the areas of the endometrium. The weakened arterial walls eventually rupture, and the blood released seeps into nearby connective tissue and destabilized capillary beds, and both the pool of blood cells and the degenerating endometrial tissues then finally break away from the uterine wall and enter the uterine center (lumen) to be eventually shed out of the female body through the cervix and vagina as menstrual flow. The process of endometrial sloughing is called menstruation, a process that is accompanied by bleeding for three to five days, over which time roughly thirty-five to fifty milliliters of blood is lost. The proliferative phase of the menstrual cycle is marked by the regeneration and thickening of the remaining thin endometrium for a week or two. This event is followed by the secretory phase, which is usually approximately two weeks in duration and is characterized by the continual thickening and vascularization (blood vessel formation) of the endometrium, with associated endometrial gland development and secretion of a glycogen-rich fluid. If a fertilized ovum has not implanted in the uterine lining by the end of the secretory phase, the uterine cycle begins with the onset of menses, marking day one of a new twenty-eight-day cycle.

While many females find this process relatively painless, painful menstruation, or dysmenorrhea, can result from uterine inflammation and contraction or from severe medication conditions involving nearby pelvic organs or tissues. However, many females experience a combination of troublesome physical and psychological symptoms seven to ten days before the start of menses (called the luteal phase in the ovarian cycle and the secretory phase in the uterine cycle) and sometimes overlapping with menses, which is termed premenstrual syndrome (PMS). Signs and symptoms usually increase in severity until the onset of menses and then dramatically disappear. Common symptoms of this mysterious malady, which results from complex physical and physiological changes occurring in the female body days before menses, include fluid retention (swelling), weight gain, skin eruptions (acne), breast engorgement and enlargement, headaches, food cravings (e.g., sweets, salty foods), fatigue, crying, insomnia, forgetfulness, dizziness, severe pelvic and abdominal pain, and an uncomfortable sensation of bloating. Such symptoms may be combined with associated psychological changes to elicit feelings of irritability, confusion, anxiety, nervous tension, mood swings, and depression, which overall may cause a recurring cycle of symptoms so severe that it affects the lifestyle and work of many females.

While the exact cause and mechanism of PMS has yet to be clearly described, changes in steroid-based sex hormone (e.g., estrogen and progesterone) concentrations in the bloodstream are suspected to be involved. For example, such hormones may be involved directly, by acting on peripheral organ systems, or indirectly, by modifying the release of neurotransmitters (chemical messengers released by nerve cells) in the central nervous system (the brain and spinal cord). However, some symptoms of PMS are not directly related to changes in the levels of these hormones. As there are currently no conclusive laboratory tests or medical procedures to accurately diagnose PMS, retaining records on the emergence of symptoms over a period of two to four months is recommended to reveal characteristic patterns. At present, PMS is simply treated for each individual at the symptom level, which may include recommendations for changes in lifestyle such as exercise, vitamin intake, and specific dietary regimens. Medication is often used for PMS, depending on the characteristics of the primary experienced symptom.

Over-the-counter (OTC) oral medications specifically marketed for the treatment of PMS symptoms can contain three active ingredients: an analgesic, a diuretic, and an antihistamine. Analgesics are drugs that relieve the feeling of pain. An example of an analgesic found in OTC oral PMS medications is acetaminophen. Acetaminophen is the least toxic member of a class of analgesic (and antipyretic [antifever]) medications known as the p-aminophenols. A major metabolite of phenacetin (the so-called coal tar analgesic) and acetanilide, acetaminophen is an effective and fast-acting analgesic that acts centrally to relieve mild to moderate pain. Acetaminophen acts to alleviate pain by effectively inhibiting the COX enzyme in the body, which normally catalyzes the synthesis of pain-producing chemicals called prostaglandins. Prostaglandins are local-acting chemical messengers that promote the redness, heat, and swelling of tissue inflammation and associated pain. Interestingly, the lining of the uterus releases prostaglandins during menses. Prostaglandins are known to stimulate uterine contractions but are inhibited in the presence of high progesterone levels. If pregnancy does not occur, progesterone levels decrease rapidly and prostaglandin production increases. The released prostaglandins then target the middle smooth muscle layer of the uterus to cause an increase in painful uterine contraction, which assists in the flow of menstrual discharge but may also lead to an increase in uterine blood vessel constriction. Such vascular constriction may cause a decrease in available oxygen to the uterine muscle, resulting in additional intense pain (e.g., cramping). Acetaminophen also specifically acts to alleviate the pain of headaches associated with PMS fever by effectively inhibiting the COX enzyme and prostaglandin production in the brain, thereby producing a pain-relieving effect by increasing the pain threshold (the conscious awareness of pain through nerve transmission at a particular stimulus intensity). Thus, acetaminophen temporarily relieves minor aches and pain attributable to headache, backache, and uterine cramping associated with PMS.

Drugs called diuretics (from the Greek word *diouretikos*, meaning promoting urine) cause an overall removal of fluid from the body by targeting the kidneys and increasing the rate of urine output and sodium excretion from the body. Diuretics such as pamabrom (8-bromo-3,7-dihydro-1,3-dimethyl-1H-purine-2,6-dione compound with 2-amino-2-methyl-1-propanol [1:1]; $C_{11}H_{18}BrN_5O_3$) are often used to combat the symptoms of bloating and fluid retention associated with PMS by causing an increase in fluid loss through urination. Diuretics have many different mechanisms of action, but all affect transport activity or water reabsorption along the nephron and collecting duct system of the kidneys. Proposed by the Food and Drug Administration as an OTC diuretic for menstrual drug products in 1988, pamabrom is therapeutically considered a drug combination, meaning a single preparation containing two active ingredients for the purpose of a concurrent administration as a fixed-dose mixture. Chemically, it contains both a propanolamine (amino alcohol) group and a theophylline group. The theophylline chemical group of pamabrom increases renal (kidney) blood flow and glomerular filtration rate (the total amount of filtrate solution formed from the blood per minute by the kidneys) and may also decrease proximal tubular reabsorption of sodium and water, overall causing a mild diuretic (water loss) effect. The amino alcohol group of pamabrom may also function as a diuretic, as alcohols act indirectly as diuretics by suppressing the release of antidiuretic hormone from the posterior pituitary gland located near the human brain. After release from the posterior pituitary gland, antidiuretic hormone normally travels through the bloodstream and targets the distal convoluted tubules and collecting ducts of the kidney nephrons to promote water retention, thus decreasing urine output. The diuretic effect of pamabrom is directly dependent on the amount of drug consumed and the duration of use. Thus, pamabrom temporarily relieves the water weight gain, bloating, swelling, and full feeling associated with PMS.

Antihistamines antagonize (inhibit) the activity of a chemical called histamine. Histamine [2-(4-imidazolyl)-ethyl-amine], a vasoactive monoamine formed by the decarboxylation of histidine by the enzyme histidine carboxylase, chemically mediates local immune responses. It is located in most body tissues but is highly concentrated in the lungs, skin, and gastrointestinal tract. In the CNS, it may serve as a neurotransmitter. Histamine acts by binding to receptors on target cells. Different cell types express different histamine receptor (H) types, which currently include H_1, H_2, H_3, and H_4. Overall, histamine contributes to the inflammatory response, acts on the smooth muscles of blood vessels to cause vasodilation (increase in blood vessel diameter), causes an increase in the permeability of blood vessel walls, and affects nearby sensory nerves, resulting in itching (also called pruritus). The effects of histamine cause the familiar symptoms of allergy, which include sneezing, inflammation of the nasal passageways, nasal itching, watery nasal discharge, and itchy, inflamed, tearing eyes. When released in the lungs, histamine causes smooth muscle contraction of the airway bronchioles, which is an attempt by the body to prevent the offensive allergens from entering the lung tissue. Unfortunately, this type of response leads to the symptoms of wheezing and shortness of breath like that experienced by individuals with the life-threatening condition called asthma.

In 1937, D. Bovet and A. Staub discovered the first H_1 receptor antagonist. This discovery marked the "first generation" of antihistamines. Antihistamines suppress the wheal (swelling) and flare (vasodilation) symptoms, typical of what is called a type I immediate hypersensitivity immune response, by blocking the binding of histamine to its receptors on nerves, vascular smooth muscle, glandular cells, endothelium tissue, and mast cells. Thus, antihistamines antagonize the action of basophils (granular white blood cells) and mast cells, which release vasoactive amines and other chemical mediators of inflammation. Classic first-generation antihistamines block the action of histamine at specific histamine receptors (e.g., H_1 receptors). First-generation antihistamines are small lipophilic molecules, so they may cause potentially adverse effects (e.g., sedation, impaired cognition, blurred vision, gastrointestinal symptoms, dryness of mouth, heart palpitations, urinary retention) because their structure closely resembles that of blockers of cholinergic (specifically muscarinic) and α-adrenergic receptors of the autonomic nervous system, and because of their ability to cross the blood-brain barrier to affect the

CNS. These H_1 receptor antagonists are reversible, competitive inhibitors of the pharmacological actions of histamine on H_1 receptors.

An example of a first-generation H_1-blocking compound available in OTC oral PMS medications is pyrilamine maleate [also called paramal or mepyramine maleate; 1,2-ethanediamine, N-((4-methoxyphenyl)methyl)-N,N'-dimethyl-N-2-pyridinyl-, (Z)-2-butenedioat; $C_{17}H_{23}N_3O{\cdot}C_4H_4O_4$)], which belongs to the ethylenediamine-derivative chemical class of first-generation H_1 receptor antagonist agents. Thus, because of the blockade of H_1 and muscarinic receptors in the nervous system and associated sedative effects, pyrilamine maleate may alleviate such PMS symptoms as insomnia and nervous tension. In addition, this antihistamine may alleviate cramping by inhibiting histamine-induced uterine smooth muscle contraction during menses. While peak plasma concentrations are usually reached two to three hours after oral administration, adverse gastrointestinal effects and sensitivity are common with ethylenediamine-derivative antihistamines in humans.

PRURITUS (ITCH) TREATMENTS

Pruritus, synonymous with itch, is often defined as an unpleasant sensation on the skin that provokes the desire to rub or scratch the affected area to obtain relief. Itch of cutaneous origin (also referred to as "pruritoceptive" or "peripheral" itch) is a very common medical symptom, consciously perceived as being a mere prickling or tingling or may be so intense as to be nearly intolerable. There are regional differences in the severity of the body's reaction to the itch, and chronic itching may lead to constant discomfort and frustration, with extreme cases causing disturbed sleep, anxiety, and depression.

Localized pruritus may result from the fact that certain parts of the body have a predilection for certain disease processes (e.g., scalp: eczema, psoriasis; eyelid: airborne irritants or allergens; nose: hay fever; fingers: eczema, scabies; legs: gravitational eczema; groin: vaginitis). General pruritus that occurs over large body areas may be the result of external causes such as exposure to severe changes in environmental climate (e.g., temperature, humidity), exposure to irritating particulate matter, allergens, or chemicals (e.g., detergents), or excessive bathing (e.g., aquagenic pruritus). General pruritus may also be the result of various skin problems (e.g., overly dry skin, dermatitis, prickly heat, acne, eczema, psoriasis, sunburn, urticaria), pregnancy, and various general systemic diseases, including microorganismal infestation (e.g., viruses, parasites, bacteria, fungi), endocrine disorders (e.g., diabetes, hypothyroidism or hyperthyroidism, hypoparathyroidism), nerve disorders (e.g., multiple sclerosis), liver malfunction (e.g., cholestasis, obstructive jaundice, cirrhosis), kidney failure, blood imbalances (e.g., iron deficiency anemia, polycythemia), autoimmune syndromes, and various types of malignant cancers. In some cases, itching can be the first sign of a serious internal illness. Itch can also be a side effect of a wide variety of drugs, including aspirin, birth control pills, testosterone, morphine, penicillin, and the antimalaria drug chloroquine. While pruritus also may be solely psychogenic or psychiatric in origin, emotional stress and psychological trauma intensify all forms of pruritus.

Sensory receptors are specialized cells or cell processes that provide the CNS (brain and spinal cord) with information concerning the inside or outside of the body. A sensory receptor detects an arriving stimulus and translates the stimulus into an electrochemical signal that can be conducted to the CNS for processing. Cutaneous itch receptors are unspecialized free nerve endings located near the dermal-epidermal junction of the skin. These nerve endings lack an electrochemical signal conduction velocity-enhancing covering called a myelin sheath and are very small in diameter, characterized as type C fibers that conduct impulses very slowly at one meter per second or less. These itch-transmitting unmyelinated C fibers enter the dorsal horn area of the spinal cord gray matter, synapse there with other secondary neurons (nerve cells), which in turn cross over to the contralateral spinothalamic tract (a collection of nerve fibers in the CNS) and ascend to the thalamus of the brain. Within the thalamus, tertiary (third-order) neurons then relay the sensation of itch to the cerebral sensory cortex. The cerebral cortex is the part of the brain involved with processing conscious awareness and allows the sensation of a low-intensity stimulus affecting the skin to be perceived as an "itch" that needs to be scratched.

Biochemicals that have been shown to be direct mediators of the physiological itch response include histamine and substance P. Histamine, produced by mast cells within the dermis, is released and causes itching by promoting the dilation of local blood vessels, increasing the permeability of local capillaries, and promotes exudate formation as a result of dermal mast cell injury. Of the two different subclasses of histamine receptors (H_1 and H_2) identified in human skin, only the H_1 receptor has been shown to be involved in histamine-induced itching. Substance P, a neurotransmitter peptide released by the unmyelinated type C fibers, dilates blood vessels and indirectly activates mast cells to release histamine, thereby causing additional redness, swelling, and itching of the skin.

As already described, pruritus is an obvious sensory feature of many different external and internal sources. The motor response that the sensation of itching evokes (i.e., scratching with the nails of the hands or feet), if not controlled, often perpetuates and intensifies the symptom and may lead to serious skin damage (e.g., abrasion, laceration, lichenification [skin thickening]), thereby reducing the ability of the skin to function as a protective barrier against infectious disease. It has been suggested that the act of scratching excites the local nerve plexus (through which the itch sensation is transmitted), which then disrupts the rhythmic systematic flow of afferent impulses traveling toward the spinal cord required for the itch sensation, alleviating the itch. However, skin damage caused by scratching itself leads to the release of itch-mediating chemicals (e.g., histamine), which once again elicits a scratch response. Thus, controlling the itch-scratch-itch cycle with home remedies, including cold compresses and soaking in tepid baths containing colloidal oatmeal, or with topical over-the-counter anti-itch products, is of paramount concern.

Topical treatments that provide symptomatic relief of pruritus can include active ingredients such as diphenhydramine HCl, pramoxine hydrochloride, camphor, menthol, phenol, dimethicone, zinc acetate, and calamine. Diphenhydramine hydrochloride [2-(diphenylmethoxy)-N,N-dimethylethylamine hydrochloride; $C_{17}H_{21}NO \cdot HCl$], an H_1 antagonist of the ethanolamine class, acts as an antihistamine by competing with free histamine for binding at H_1 receptor sites. Blockage of such sites also suppresses the formation of edema, flare, and pruritus that normally result from histaminic activity. Topical diphenhydramine provides temporary relief from pruritic skin irritation, possibly attributable to an anesthetic (numbing) effect resulting from the decreased permeability of nerve cell membranes to sodium ions, thereby preventing the transmission of nerve impulses. Pramoxine hydrochloride (morpholine 4-[3-(4-butoxyphenoxy)propyl]-, hydrochloride) is also a local anesthetic that interferes with the function of nerves that sense pain and thus temporarily relieves minor pain, itching, and discomfort. In addition, both camphor ($C_{10}H_{16}O$), a ketone synthesized from oil of turpentine ingredients or obtained naturally by steam distillation of the wood of the camphor tree (*Cinnamomum camphora*), and menthol ($C_{10}H_{19}OH$), a stearopten obtained from oil of peppermint (*Mentha piperita*), also exhibit anesthetic properties by producing sensations of warmth, followed by a chilling sensation. Phenol (C_6H_5OH), an aromatic alcohol obtained from coal tar, acts as an antiseptic and disinfectant to decrease the chance of wound infection. Dimethicone ($[-Si(CH_3)_2O-]_n$) imparts lubricity and softness to the antipruritic cream product and serves as a skin protectant, while both zinc acetate [$(CH_3COO)_2Zn2H_2O$] and calamine [$(ZnOH)_2SiO_3$] act as skin protectants and drying agents.

SPERMICIDE (BIRTH CONTROL PRODUCT)

The human reproductive system is the only organ system in both males and females that is not essential to the life of the individual. Although the reproductive systems of the human male and female possess many

differences, their common purpose is to produce offspring and thus per-
petuate the species. In general, the role of the human reproductive system is to synthesize, maintain, nourish, and transport functional male and
female reproductive cells called gametes. The reproductive system includes
gonads, or reproductive organs that produce the gametes and secrete
steroid-based sex hormones; ducts that receive and transport the gametes;
accessory glands and organs that secrete fluids into the ducts or other
accessory structures to allow for adequate lubrication and gamete functioning; and perineal structures collectively called external genitalia that
enhance the human sexual response and allow for efficient reproductive
functioning during a copulation event between males and females. Male
gametes, called sperm, are produced in gonads called the testes, and female
gametes, called ova (eggs), are produced in gonads called the ovaries.
When a male and a female gamete unite, fertilization, also known as con-
ception, occurs. In sexual reproduction, fertilization results in the formation of a completely new and genetically unique organism. Specifically,
the single cell resulting from the union and fusion of the male and female
gametes, called a zygote, contains a mixture of chromosomes (DNA) from
the two parents. Through a development process of repeated mitotic cell
divisions and differentiation (specialization), the zygote undergoes gradual transformation and maturation within the female uterus into a complex
multicellular new human being.

Sperm are specialized cells produced in the male testes, which are paired
oval glands measuring approximately five centimeters (two inches) in length
and two and a half centimeters (one inch) in diameter that are supported
externally by a cutaneous outpouching of the male abdomen consisting
of loose skin and connective tissue called the scrotum. Within the testes,
septal tissue extensions divide the testes internally into 200 to 300 wedge-
shaped compartments called lobules. Each lobule contains one to four
tightly coiled tubules called seminiferous tubules. Within the seminiferous
tubules, sperm are produced by a unique cellular division process called
spermatogenesis. Controlled by hormones including testosterone, the
process of spermatogenesis begins during puberty within males and normally continues throughout life. A healthy adult human male produces
approximately 400 million sperm daily.

A sperm is a highly adapted and mobile cell capable of reaching and
penetrating a female ovum (egg). Each sperm has three distinct regions:
the head (genetic region), the midpiece (metabolic region), and the tail
(locomotor region). The head is a flattened ellipse containing a nucleus
with densely packed chromosomes (DNA) and a dense membranous
granule (actually a specialized lysosome organelle) at the tip called the
acrosomal cap that contains enzymes, including proteinases and hyalu-
ronidase, which aid in the penetration of the sperm cell into the ovum. A

short neck attaches the head to the midpiece, a region containing organelles called mitochondria that absorb nutrients from the surrounding fluid (called semen) and produce energy (in the form of adenosine triphosphate) for sperm locomotion. The tail consists of an organelle called a flagellum, which allows the sperm to propel through the female reproductive tract in a unique whiplike corkscrew motion.

Contraception (*contra* = against, *cept* = taking) and birth control are synonymous terms. While many individuals use various birth control strategies during their reproductive years, readily available OTC chemical methods of contraception containing spermicide chemicals are frequently chosen. Various foams, creams, jellies, vaginal suppositories, sponges, and douches that contain spermicidal agents make the female vagina and cervix unfavorable for sperm survival and thus decrease the likelihood of a successful fertilization event during intercourse. Once the sperm-killing chemicals are inserted into the female vagina to coat the vaginal surfaces and cervical opening into the uterus, relative contraception protection is provided for approximately one hour. Spermicides can be effective when used alone but are significantly more effective when used with physical barrier devices, including condoms, vaginal pouches, diaphragms, cervical caps, or sponges.

For approximately thirty years, one of the most widely used Food and Drug Administration-approved active ingredient spermicides in various OTC contraceptive products has been nonoxynol-9 (N-9). N-9 is an almost colorless liquid nonionic detergent (surfactant, a surface-active agent) that inactivates sperm via disruption and disaggregation of the outer protective plasma membrane. Thus, the number of active and viable sperm decreases significantly. This activity stems from the ability of nonionic surfactants, which lack a specific charge and possess a hydrophilic head region and a hydrophobic tail region, to dissolve lipid-containing membranes. A nonoxynol [a-(4-nonylphenyl)-w-hydroxypoly(oxy-1,2-ethanediyl) or polyethyleneglycol mono(nonylphenyl) ether] is a nonionic surfactant mixture prepared by reacting nonylphenol with ethylene oxide. The hydrophilic (or water-soluble) portion of the molecule contains the polyethylene oxide group. The average number of ethylene oxide units (n) per molecule is indicated by the number after nonoxynol (e.g., nonoxynol-9 for n = 9). A multicomponent mixture of oligomers (at least seventeen are known as characterized by high-performance liquid chromatography) in the commercially available form, N-9 is used as a spermicidal ingredient in spermicidal lubricants, both preapplied to spermicidal condoms and packaged separately.

TOPICAL ANTIBIOTIC TREATMENTS

Antibiotics are typically described as antimicrobial agents of natural origin that are produced by microorganisms, which elicit a lethal or growth-inhibitory effect on a range of other types of microorganisms. These agents are molecules produced as secondary metabolites mainly by microorganisms inhabiting the soil, including molds and bacteria. In most cases, antibiotic production seems to be related to the sporulation process of the organismal life cycle. While many hundreds of different compounds possessing antibiotic activity have been identified from microorganisms since the early twentieth century, only a few compounds isolated have been shown to be both clinically therapeutic in the treatment of infectious disease and minimally toxic after administration. The modern understanding of antibiotics as chemically therapeutic agents started with A. Fleming's credited 1929 discovery of a fungal (common bread mold) metabolite from *Penicillium notatum* that demonstrated potent bactericidal effects. Termed penicillin, this antibiotic was not isolated and purified until the period of World War II (1939–1945), when two scientists, H. Florey and E. Chain, managed to produce penicillin on an industrial scale for widespread use. In addition, by the 1950s, several other antibiotics were in clinical use as the result of G. Domagk's 1935 discovery of synthetic chemicals (sulfonamides) with broad antimicrobial activity, along with intensive research concerning other antimicrobial agents of natural origin. Three different antibiotics used in varying concentrations in present-day over-the-counter antibiotic treatments are neomycin, bacitracin, and polymyxin B. These antibiotics are usually formulated within chemically inactive creams or petrolatum-based ointments for topical application.

The range of bacteria or other microorganisms that are affected by a particular antibiotic is termed its spectrum of action. Broad-spectrum antibiotics are those that kill or inhibit a wide range of Gram-positive and Gram-negative bacteria, whereas narrow-spectrum antibiotics are mainly effective against either Gram-negative or Gram-positive bacteria. The Gram stain, named after its developer, C. Gram, is a laboratory staining technique that distinguishes between two groups of bacteria by the identification of differences in the structure of their cell walls. While the cell membrane is the critical barrier of the bacterial cell, separating the inside ribosome and nucleic acid components from the outside of the cell, the cell wall is a rigid structure outside the cell membrane that provides support and protection. Although the cell walls of these bacteria are similar in chemical composition, Gram-positive bacteria remain colored after the staining procedure, whereas Gram-negative bacteria do not retain dye. The cell wall of Gram-negative bacteria consists of a thin layer positioned between an outer lipid-containing cell envelope and an inner cell membrane, whereas the Gram-positive cell wall is much thicker, lacks the cell envelope, and contains additional substances such as teichoic acid (polymers composed of glycerol or ribitol).

Many antibacterial agents produce a clinically beneficial effect by interfering with bacterial cell wall synthesis, cell membrane physiology, or protein synthesis. Peptidoglycan (and its synthesis pathway) is a major component of bacterial cell walls and thus one of the major targets of antibiotics in both Gram-negative and Gram-positive bacteria. Eukaryotic cells (cells that contain a true nucleus) within organisms such as humans lack cell walls and peptidoglycan. Other antibiotic compounds may target bacterial protein synthesis because bacterial ribosomes (termed 70S ribosomes) are different from the ribosomes (80S) of humans and other eukaryotic organisms. Thus, antibiotics may exert selective toxicity against bacterial pathogens without deleteriously affecting the consumer taking the antibiotic drug.

The actinomycetes are a large group of soil-inhabiting filamentous bacteria that produce many different chemical classes of antibiotics, including chemicals known as aminoglycosides. These compounds are water-soluble weak bases that are characterized by the presence of an aminocyclitol ring linked by glycosidic bonds to amino sugars in their structure. Within the actinomycetes, the species *Streptomyces fradiae* produces the broad-spectrum antibiotic neomycin, which causes the premature termination of protein synthesis (translation) in bacteria. Specifically, the aminoglycosides irreversibly bind to the 30S subunit of the bacterial ribosome and interfere with the formation of the initiation complex to cause misreading of the mitochondrial RNA. First isolated by S. Waksman from a strain of *S. fradiae* in 1949, neomycin is mainly used topically in the treatment of skin and mucus membrane infections, wounds, and burns. Although used systemically in some cases, it is highly toxic. Neomycin sulfate, often used in topical antibiotic treatments in lieu of neomycin, is the sulfate salt of neomycin B and C, which are also produced by the growth of *S. fradiae*.

Endospore-forming *Bacillus* bacterial species produce a chemical class of antibiotics known as polypeptides, including bacitracin and polymyxin B. Polypeptide antibiotics consist of an amino acid chain. Bacitracin ($C_{66}H_{103}N_{17}O_{16}S$), a metal-dependent polypeptide complex of closely related analogues produced by common soil and water bacteria, including *Bacillus subtilis* and *Bacillus licheniformis*, is a narrow-spectrum antibiotic effective against Gram-positive bacteria. It prevents cell wall growth by inhibiting the release of the muropeptide monomer subunits of peptidoglycan from the undecaprenyl pyrophosphate lipid carrier molecule that carries the subunit to the outside of the bacterial cell membrane. Binding to the undecaprenyl pyrophosphate lipid carrier impedes the dephosphorylation of the lipid carrier, which then obstructs the regeneration of undecaprenyl phosphate, thus preventing recycling of the bacterial transport system and the cell wall synthesis mechanism. Synthesis of teichoic acid, a key cell wall component in Gram-positive bacteria, which requires the same carrier molecule, is also inhibited. Bacitracin is limited to topical applications because it also interferes with sterol synthesis in mammalian cells by binding to pyrophosphate intermediates, thereby eliciting a toxic response when used systemically in humans.

Polymyxin B, a naturally occurring cyclic decapeptide produced by *Bacillus polymyxa*, is a narrow-spectrum antibiotic effective against Gram-negative bacteria and is usually limited to topical applications. It consists of a seven-member ring containing four diaminobutyric acid (Dab) residues, one threonine residue, and a hydrophobic segment (i.e., dPhe-Leu and a linear amino-terminal region composed of three amino acids, Dab-Thr-Dab, together with an eight- or nine-carbon fatty acid [6-methyl heptanoic acid and octanoic acid, respectively] forming a long hydrophobic tail). Considered one of the most efficient cell-permeabilizing compounds, it binds to membrane phospholipids within the cell membrane of the bacterium and thereby interferes with membrane physiological function. Polymyxin B acts a detergent, increasing the permeability of the membrane to cause the contents of the bacterial cell to leak out. Because of similarities in the phospholipid chemical composition of bacterial and eukaryotic cell membranes, polymyxin B is rarely used as a systemic antibiotic treatment in humans, as it does not specifically target and destroy only bacterial cell membranes. Polymyxin B sulfate, frequently used in topical antibiotic treatments in lieu of polymyxin B, is the sulfate salt of polymyxin B_1 and B_2, which are also produced by the growth of *B. polymyxa*.

Topical antibiotic treatments typically also include a local anesthetic (numbing agent), such as pramoxine hydrochloride, that interferes with the function of nerves that sense pain.

TOPICAL ANTI-INFLAMMATORY TREATMENTS

Inflammation is generally considered a localized physiological response to injury or pathogen infection of the body tissues. Inflammation is categorized as a type of nonspecific response to tissue injury from physical trauma (e.g., impact, abrasion, distortion), intense heat, or irritating chemicals. The common factor among these triggers of inflammation is a cascade involving cell death, connective tissue fiber damage, and/or general tissue injury. In addition, tissue injury may allow for the introduction of foreign proteins or pathogens (disease-causing agents) such as viruses, fungi, and bacteria. Overall, the changes within the interstitial biological environment lead to a complex process called the inflammatory response. The purpose of this response includes temporary repair of the damaged tissue at the injury site and prevention of additional pathogen entry, prevention (or slowing) of the spread of pathogenic agents to other body areas, disposal of cell debris and pathogens, and mobilization of systemic immune defenses to allow for regeneration and wound healing. Such a process causes an increase in vasodilation (dilation of blood vessels) and interstitial fluid accumulation at the injury site, an alteration in the chemical composition of interstitial tissue fluid, and the release of chemical signals (e.g., prostaglandins and histamine), protein factors (e.g., heparin, kinins, complement, and lymphokines), and potassium ions. For example, histamine (released from cells called mast cells and basophils) promotes the vasodilation of local arterioles, increases the permeability of local capillaries, and promotes exudate formation, whereas prostaglandins (fatty acid molecules produced from arachidonic acid and located in all cell membranes) sensitize blood vessels to the effects of other inflammatory mediators, induce pain, and allow for the generation of pain- and inflammation-causing free radicals.

Four key medical signs of acute (short-term) inflammation at an anatomical body site are redness and heat (attributable to the increased blood flow and blood volume), swelling, and the sensation of pain (attributable to the increased presence of interstitial fluid exudates and edema-causing adjacent nerve ending stimulation). Many over-the-counter products used as topical anti-inflammatory treatments typically provide temporary relief from minor skin inflammation, itching, and superficial rash caused by medical conditions such as eczema, psoriasis, hives (urticaria), seborrheic dermatitis, and diaper rash and from skin contact with potential allergic reaction-causing environmental factors such as soaps, detergents, poison ivy, poison oak, poison sumac, insect bites, jewelry, and cosmetics. These topical treatments typically contain the compound hydrocortisone (i.e., cortisol; $C_{21}H_{30}O_5$), which is a steroid hormone produced by the cortex of the adrenal (suprarenal) gland that is naturally secreted in a characteristic diurnal rhythm and is manufactured synthetically for medical use. Topical corticosteroids (including hydrocortisone) share potent combined anti-inflammatory, antipruritic (relieving and/or preventing itching), and vasoconstrictive (constriction of blood vessels) actions. Hydrocortisone inhibits inflammation and pain by stabilizing lysosomal membranes and preventing vasodilation, potentially acts as an antioxidant to counteract the production of free radicals, relieves the sensation of itching by inhibiting the release of histamine, and diminishes redness and swelling by enhancing the vasoconstrictive effects of the hormone epinephrine. These combined effects cause an overall depression of the local inflammatory response cascade to provide temporary comfort while also causing a delay in wound healing.

The extent of topical corticosteroid absorption through the skin and the subsequent therapeutic effect is determined by many factors, including the chemical vehicle (i.e., the chemical consistency of the ointment or cream product containing the hydrocortisone), the integrity of the epidermal barrier (i.e., applying the treatment on normal healthy skin versus diseased or inflamed skin), and the use of protective and occlusive barrier dressings on the skin. Once absorbed through the skin, hydrocortisone typically binds to plasma proteins within the bloodstream, is metabolized primarily by the liver, and is excreted from the body by the kidneys.

Some topical over-the-counter creams and ointments contain combinations of corticosteroids and antibiotics (e.g., bacitracin zinc, polymyxin B sulfate, neomycin sulfate) and are used to treat ear, eye, and skin infections caused by the presence of bacteria. In addition, these products may also contain lubricants (e.g., glycerin, white petroleum, beeswax, light mineral oil), emulsifiers (e.g., glyceryl stearate, PEG-40 stearate, polysorbate 60, ceteareth-20), emollients (e.g., isostearyl neopentanoate, cetyl palmitate), humectants (e.g., stearyl alcohol, purified water), pH-regulating agents (e.g., calcium acetate, sodium citrate), preservatives (e.g., citric acid, sorbic acid, methylparaben, propylparaben), surfactants (e.g., stearyl alcohol, sodium lauryl sulfate), vitamins (e.g., tocopheryl acetate [vitamin E], vitamin A palmitate, vitamin D), solution binders (e.g., maltodextrin), skin-soothing natural anti-itch products (e.g., *Aloe barbadensis* leaf juice or gel, *Avena sativa* [oat] kernel flour), and agents that enhance hydrocortisone drug permeation through the skin (e.g., isopropyl myristate).

VITAMINS

Within complex biological organisms, nutrients such as proteins, carbohydrates, and fats serve as building blocks and combine with various other substances to yield energy and maintain cellular functioning. These chemical reactions are catalyzed (accelerated) by individual enzymes, which are located in specific body regions. Vitamins are potent organic (carbon-based) compounds that mainly function as coenzymes (or parts of coenzymes) that individually act in concert with each enzyme to accomplish a specific type of reaction catalysis process. Vitamins are classified as either fat-soluble or water-soluble. Fat-soluble vitamins (e.g., A, D, E, and K) bind to ingested fats and are absorbed into the body along with their digestive products. Except for vitamin K, fat-soluble vitamins are generally stored within the body. Water-soluble vitamins (e.g., B complex and C) are absorbed along with water from the gastrointestinal tract, and unless metabolically used, they are usually excreted in the urine. A noted exception regarding absorption is water-soluble vitamin B_{12}, which must first bind to a chemical called intrinsic factor, which is produced by the stomach, to be absorbed into the body.

These essential organic compounds are biologically required only in small amounts, but deficiency tends to result in a diseased state for the individual. With a few noted exceptions (e.g., vitamins D, K, and B_3), most vitamins are not manufactured within the body and thus must be obtained via food sources or direct vitamin supplementation. The value of certain foods in maintaining health was recognized long before the first vitamins were actually isolated and characterized. Nearly 3,500 years ago, for example, Egyptians recognized that night blindness (caused by vitamin A deficiency) could be treated with specific foods. In the eighteenth century, it was demonstrated that the addition of citrus fruits to the diet could prevent the development of scurvy (caused by vitamin C deficiency). In the nineteenth century, it was shown that substituting unpolished for polished rice in a rice-based diet could prevent the development of beriberi (caused by vitamin B_1 deficiency). In 1906, British biochemist F. Hopkins demonstrated that foods contained necessary "accessory factors" in addition to proteins, carbohydrates, minerals, and water. The word "vitamin" became a modern vocabulary term as shortened from the original word *vitamine*, as used by Polish chemist C. Funk in 1912 to describe the "vital amine" (a compound containing a nitrogen bound to three hydrogen atoms [$-NH_3$] that is vital to our health) antiberiberi accessory growth factor substance then discovered in unpolished rice. The term "vitamin" soon came to be applied to all accessory growth factors in general when many scientists identified, purified, and synthesized the thirteen vitamins and discovered that not all of the factors contained the nitrogen-based chemical amine groups. Vitamins were originally categorized based on their body function and assigned letter names to simplify discussion. As their chemical structures were determined, chemical names were also used.

Human vitamin requirements are generally expressed in terms of the recommended dietary allowance (RDA). These RDA values, as established by the Food and Nutrition Board of the National Academy of Sciences/National Research Council in the United States and by the Food and Agriculture Organization and World Health Organization for different worldwide population groups, represent the amount of essential nutrients that, if acquired daily, are considered sufficient to meet the known nutritional requirements of most healthy individuals within the population. In some cases, the average requirements are not known with precision; RDA values are then based on average dietary intake within a population, plus extra as a margin of safety to account for increased demands (e.g., during illnesses, etc.). While once expressed in terms of international units, the strength of a vitamin or the amount of a vitamin necessary to produce a certain biological effect is currently expressed directly in micrograms or milligrams (metric weights). Many vitamins work together to regulate several body processes, and either an overabundance or insufficient amount of vitamins may potentially disturb the internal balance, or homeostasis, within the body and potentially lead to a disease state or, in some cases, death.

Fat-soluble vitamins, including A, D, E, and K, seem to have highly specialized functions. The intestine absorbs fat-soluble vitamins, and the lymphatic system transports these vitamins to various body regions. Because fat-soluble vitamins easily dissolve in lipids (fats), hydrocarbons, and similar solvents, they normally diffuse through the cell membranes and into other lipids of the body, including adipose (fat) tissue and the lipid inclusions within the liver. Thus, the body usually maintains a significant storage reserve of these vitamins, and normal metabolism may persist for quite a long time (usually several months) after dietary sources of these vitamins have been excluded.

Vitamin A

In 1913, T. Osborne and L. Mendel, while conducting experiments using rats, discovered that butter contained a growth-promoting, fat-soluble nutrient necessary for development. Soon known as vitamin A, its chemical character was established in 1933, and it was first synthesized in 1947. Vitamin A consists of three biologically active molecules: retinol [3,7-dimethyl-9-(2,6,6-trimethyl-1-cyclohexen-1-yl)-2,4,6,8-nonatetraen-1-ol; $C_{20}H_{30}O$], retinal (retinaldehyde), and retinoic acid. Each of these compounds is derived from the plant precursor molecule, β-carotene (a member of a family of molecules known as carotenoids). β-Carotene, which consists of two molecules of retinal linked at their aldehyde ends, is also referred to as the provitamin form of vitamin A. Ultraviolet light inactivates vitamin A, so it is often destroyed via oxidation upon exposure to heat, light, or air. Within the intestine, ingested β-carotene is cleaved via enzymes to initially yield retinal and then subsequently reduced via enzymes to retinol. Retinol is esterified, delivered to the blood, and then delivered to the liver for storage as a lipid ester. Thus, nearly 90 percent of vitamin A within the body is contained within the liver. Dietary sources of vitamin A include the provitamin precursor carotene, found in carrots, deep-yellow and deep-green leafy vegetables, and vitamin A, found in fish liver oils, egg yolk, liver, whole milk, butter, and cheese. Vitamin A has three major roles within the body: gene regulation, proper visual functioning, and general body organ health and maintenance. Retinol and retinoic acid are considered hormones of the steroid/thyroid hormone superfamily of proteins. Within cells, both retinol and retinoic acid bind to specific receptor proteins, and this receptor-vitamin complex then interacts with several genes involved in growth and differentiation to affect their expression. For example, gene expression patterns involved with early processes of embryological development, including organogenesis and limb development, are affected by retinoic acid. Vitamin A has a direct role in the process of vision, as the photosensitive compound of most mammalian eyes is a protein called opsin (present in the photoreceptor rod cells within the retina of the eye), which is covalently coupled to an aldehyde form of vitamin A (*cis*-retinal). Exposure of the eye to ultraviolet light results in a series of reactions within the photoreceptor cells that are mediated by photon absorption by *cis*-retinal, which eventually leads to propagation of nerve impulses from the optic nerve of the eye to the brain, where vision is processed. Retinol also functions in the synthesis of certain glycoproteins and polysaccharides necessary for mucus production and normal growth regulation within most body organ systems. For example, vitamin A is required to maintain the integrity of the skin and mucus membranes, normal bone and tooth development, and normal reproductive capabilities and additionally acts as an antioxidant to provide anticancer and antiatherosclerosis effects.

Vitamin D

While researching a solution to the bone disease called rickets in 1922, E. Mellanby discovered vitamin D. In the United States, the enrichment of milk with this vitamin was, and continues to be, extremely effective against rickets, which is the faulty mineralization of bones and teeth in growing young children, producing soft bones and conditions known as "bowlegs" and "knockknees." Normally, a vitamin is defined as a substance that is essential for the maintenance of life-sustaining metabolic body processes yet is not synthesized by the body on its own. The single exception to this rule is vitamin D, which can be synthesized in the skin, but only when exposed to direct sunlight (ultraviolet light). Vitamin D refers to a group of chemically distinct steroids that exhibit qualitatively the biological activity of calciol [also called cholecalciferol; vitamin D_3; $(3\beta,5Z,7E)$-9,10-secocholesta-5,7,10(19)-trien-3-ol; $C_{27}H_{44}O$]. Stable when exposed to heat, light, acids, alkalis, and oxidation, vitamin D is a fat-soluble compound concentrated in the liver, skin, and kidneys that is essential for calcium and phosphorus metabolism in animals. Vitamin D is therefore important for normal mineralization of bone and cartilage, neuromuscular functioning, tooth formation, and blood clotting. The biologically active form of the hormone is 1α,25-dihydroxycholecalciferol, also called calcitriol. Active calcitriol is derived from ergosterol (produced in plants) and from 7-dehydrocholesterol (produced in the skin). After ultraviolet irradiation, ergosterol is converted to ergocalciferol (vitamin D_2) in plants, and 7-dehydrocholesterol is converted to chole-calciferol (vitamin D_3) in the skin. The same enzymatic pathways in the body then process vitamins D_2 and D_3 to D_2-calcitriol and D_3-calcitriol, respectively. Cholecalciferol (or ergocalciferol) is then absorbed from the intestine and transported to the liver for further chemical modification. For example, in the liver, cholecalciferol is hydroxylated to form 25-hydroxy-D_3, which is the major blood-circulating form of vitamin D in the body. Subsequent conversion of 25-hydroxy-D_3 to calcitriol (its bio-logically active form) occurs in the kidneys and bone tissues. Calcitriol functions in concert with both parathyroid hormone and the hormone calcitonin to regulate blood serum calcium and phosphorus levels. Major dietary sources of vitamin D include fish liver oils, egg yolk, butter, and fortified milk.

Vitamin E

In 1922, H. Evans and K. Bishop discovered vitamin E in green leafy vegetables and wheat germ while performing growth experiments with laboratory rats. Vitamin E refers to a mixture of several related fat-soluble compounds known as tocopherols. Vitamin E is a generic descriptor for all tocol [2-methyl-2-(4,8,12-trimethyltridecyl)chroman-6-ol(I, $R^1 = R^2 = R^3 = H$)] and tocotrienol derivatives exhibiting qualitatively the biologi-cal activity of α-tocopherol. Chemically related to sex hormones, vitamin E is stored primarily in muscle and adipose (fat) tissue, and to a lesser ex-tent the reproductive organs, within the body. Found largely in plant materials, including wheat germ, vegetable oils, nuts, seeds, whole grains, and dark green leafy vegetables, this vitamin tends to be resistant to heat, light, and acid exposure but is unstable in the presence of oxygen. Other sources of vitamin E include meats, milk, and eggs. The α-tocopherol [2,5,7,8-tetramethyl-2-(4',8',12'-trimethyltridecyl)-6-chromanol; $C_{29}H_{50}O_2$] molecule is the most potent of the tocopherols. Vitamin E is absorbed from the intestines and subsequently delivered to the tissues and liver, and it easily accumulates in cellular membranes, fat deposits, and other lipoproteins within the bloodstream. Its major function is to act as an an-tioxidant in preventing the peroxidation of polyunsaturated membrane fatty acids and cholesterol by scavenging free radicals and molecular oxy-gen. Thus, vitamin E, along with the antioxidant capabilities of vitamin C, assists in preventing oxidative damage to cell membranes and athero-sclerosis. Vitamin E also has been used to promote fertility.

Vitamin K

Vitamin K, which is important to blood clotting (blood coagulation), was discovered in the 1930s after an outbreak of fatal bleeding among cattle that were fed on rotten sweetclover hay, which is deficient in the vitamin. Vitamin K refers to a number of fat-soluble related compounds known as naphthoquinone derivatives that are stored in relatively small quantities within the liver and required for the bioactivation of proteins involved in homeostasis (blood clotting). The designation "K" was derived from the German *Koagulationsvitamin*. Vitamin K fat-soluble compounds are classified into three groups. The K vitamins exist naturally as K_1 (phylloquinone; 2-methyl-3-phytyl-1,4-naphthoquinone; $C_{31}H_{46}O_2$) in green plants and K_2 (menaquinones; 2-methyl-3-*all-trans*-polyprenyl-1,4-naphthoquinones) produced by intestinal bacteria. Vitamin K_3 is a synthetic menadione (2-methyl-1,4-naphthoquinone; $C_{11}H_8O_2$) that is alkylated to one of the vitamin K_2 forms of menaquinone when administered into the body. While bacteria within the large intestine of the body synthesize vitamin K, dietary sources of vitamin K include leafy green vegetables, broccoli, cabbage, cauliflower, and pork liver. Naturally occurring vitamin K_2 is absorbed from the intestines in the presence of bile salts and other lipids. Interestingly, vitamin K activity is antagonized by certain anticoagulants (e.g., warfarin) and by antibiotics that interfere with the vitamin-synthesizing capability of intestinal bacteria. As an intermediate in the electron transport chain, vitamin K participates in the oxidative phosphorylation process within all body cells and assists in the synthesis of particular proteins within the liver. However, the major function of vitamin K is in the maintenance of normal levels of blood-clotting proteins (e.g., prothrombin), which are synthesized in the liver as inactive precursor proteins. Conversion from an inactive to an active clotting factor requires the chemical modification of glutamate (an amino acid) residues via an enzymatic carboxylation reaction that requires vitamin K as a cofactor. The active clotting factors then act as biochemical intermediates in the formation of a cross-linked fibrin polymer clot to facilitate a decrease in bleeding after blood loss associated with injury to a blood vessel.

The water-soluble vitamins, including B complex and C, are absorbed along with water from the gastrointestinal tract and carried by the circulatory system to specific tissue cells, where they are used metabolically (e.g., acting as components of coenzymes). A noted exception is vitamin B_{12}, which must first bind to an intrinsic factor produced by the stomach to be absorbed. With the exception of vitamins C, folacin (folic acid), and B_{12}, insignificant amounts of these vitamins are stored within the body, and unless used, they are excreted in the urine. Because an appreciable supply of water-soluble vitamins is not maintained within the body over long periods, a daily supply is usually essential to prevent depletion. Except in the case of ingesting large or mega doses of vitamin supplements, few disease conditions are characterized as resulting from ingesting excessive levels of these vitamins.

B Complex

In 1936, a growth-promoting factor termed vitamin B was isolated from bovine (cow) milk. There are now several different types of vitamin B known and chemically characterized, and they are collectively described as B complex vitamins because of relative similarities in their properties, physiological functions, and distribution in natural resources. Mostly recognized as coenzymes, the eight B complex vitamins currently include B_1 (thiamine), B_2 (riboflavin), niacin (nicotinamide), B_6 (pyridoxine), pantothenic acid, biotin, B_{12} (cyanocobalamin), and folacin (folic acid).

In 1905, W. Fletcher discovered that there were "special nutrients" within the husks of unpolished (nonwhite) rice that prevented the occurrence of a disease called beriberi, a condition characterized by anemia, paralysis, muscular atrophy and weakness, and spasms of the muscles and legs. However, it was not until 1925 that R. R. Williams isolated and characterized this nutrient, now known as vitamin B_1. The first vitamin to be isolated, vitamin B_1 or thiamine (3-[(4-amino-2-methyl-5-pyrimidinyl) methyl]-5-(2-hydroxyethyl)-4-methylthiazolium chloride; $C_{12}H_{17}ClN_4OS$) is derived from a substituted pyrimidine and a thiazole, which are coupled by a methylene bridge. Within the brain and liver, thiamine is converted to its active form, thiamine pyrophosphate, by specific enzymes. It is a water-soluble coenzyme that is rapidly destroyed by heat and found in dietary sources such as lean meats, liver, fish, egg yolks, milk, the germ and husks of whole grains, corn, rice, leafy green vegetables, legumes, yeast, and nuts. Essential for carbohydrate metabolism, thiamine is part of the coenzyme cocarboxylase and assists the body in converting complex carbohydrates into simple sugar (glucose) molecules. Converted within the body to thiamine diphosphate (a coenzyme in the decarboxylation [release of –COOH] of α-keto acids), thiamine is required for the transformation of pyruvic acid to acetyl coenzyme A within the cellular energy cycle. Thiamine is also required for the synthesis of pentose sugars, for the oxidation of alcohol, and for the proper functioning of the nervous system, where it serves as a coenzyme in the production of the neurotransmitter called acetylcholine.

Named for its similarity to ribose sugar, vitamin B_2 or riboflavin [7,8-dimethyl-10-(d-*ribo*-2,3,4,5-tetrahydroxypentyl)isoalloxazine; $C_{17}H_{20}N_4O_6$] is a water-soluble vitamin that possesses green-yellow fluorescence and is quickly decomposed by exposure to both ultraviolet and visible light as well as alkaline chemicals. Any excess of this vitamin within the body is excreted in the urine, although small storage amounts within the liver and kidneys are carefully guarded. While this vitamin is readily available in dietary sources including liver, yeast, egg white, whole grains, meat, poultry, fish, legumes, leafy green vegetables, and milk, M. Tishler is credited with developing techniques for synthesizing this vitamin in the labora-tory. Riboflavin plays an important role in the breakdown of carbohy-drates, fats, and proteins by acting as a component of amino acid oxidase enzymes and as the precursor for coenzymes including flavin mononu-cleotide (FMN) and flavin adenine dinucleotide (FAD). Because both FMN and FAD act as hydrogen acceptors, these two bioactive forms of riboflavin occurring in tissues and cells act as coenzymes for oxidation-reduction reactions throughout the body. Riboflavin is also significant in the healthy maintenance of skin and mucus membranes, in insulating nerve cell sheath material, and in the cornea of the eye.

Vitamin B_3 or niacin (nicotinic acid; nicotinamide; 3-pyridinecarboxy-lic acid; $C_6H_5NO_2$) is a relatively noncomplex organic compound that is derived from the amino acid tryptophan and is stable upon exposure to extreme heat, light, alkalis, acids, and oxidation. Although water-soluble niacin itself can be synthesized from tryptophan in small amounts within the liver, it is available from major protein dietary sources including poul-try, meat, and fish and minor sources including liver, yeast, peanuts, po-tatoes, and leafy green vegetables. Niacin is required for the synthesis of the active forms of vitamin B_3 within the body, which include the coen-zymes nicotinamide adenine dinucleotide (NAD^+) and nicotinamide ad-enine dinucleotide phosphate ($NADP^+$). Both NAD^+ and $NADP^+$ function as coenzymes for numerous dehydrogenase enzymes and are involved in glycolysis, oxidative phosphorylation, and fat breakdown. Niacin inhibits cholesterol synthesis and acts as a peripheral vasodilator (increases blood vessel diameter). Overall, niacin is essential for the metabolism of food nutrients and the maintenance of healthy skin, nerve function, and gas-trointestinal tract function. Severe niacin deficiency is associated with a disease called pellagra, which is characterized by three major symptoms: depression, dermatitis, and diarrhea.

Isolated from liver by R. J. Williams and colleagues in 1938, pan-tothenic acid, also known as vitamin B_5 [(R)-N-(2,4-dihydroxy-3,3-dimethyl-1-oxobutyl)-β-alanine; $C_9H_{17}NO_5$], is a water-soluble and quite stable vitamin that is formed from β-alanine and pantoic acid and is stored in relatively large amounts within tissues such as the liver, kidneys, brain, adrenal glands, and heart. As the term "pantothenic" is derived from the Greek word *panthos*, meaning everywhere, this vitamin is widely distributed in dietary sources, including animal food products, whole grains, legumes, liver, rice, molasses, yeast, egg yolk, and meats. Enteric bacteria that reside within the intestines also produce some of this vita-min. Used in the breakdown of carbohydrates via the tricarboxylic acid cellular energy cycle, lipids, and some amino acids, pantothenic acid is required for the formation of coenzyme A in reactions that remove or transfer acetyl groups (e.g., the formation of acetyl coenzyme A from pyruvic acid; the oxidation and synthesis of fatty acids) and is a component of the acyl carrier protein domain of the enzyme fatty acid synthase. In addition, at least 70 enzymes have been identified as requiring coenzyme A or acyl carrier protein derivatives for their function. This vitamin is also involved in the synthesis of various steroid hormones and of the nitrogen-based heme group of hemoglobin, which is an oxygen-carrying molecule located within human red blood cells.

The first naturally occurring form isolated by numerous scientists in 1938, and structurally confirmed by chemical synthesis in 1939, vitamin B_6 is currently recognized as a group of three pyridines occurring in both free and phosphorylated forms in the body: pyridoxine (pyridoxol; 5-hydroxy-6-methyl-3,4-pyridinedimethanol; $C_8H_{11}NO_3$), pyridoxal [3-hydroxy-5-(hydroxymethyl)-2-methyl-4-pyridinecarboxaldehyde; $C_8H_9NO_3$], and pyridoxamine [4-(aminomethyl)-5-hydroxy-6-methyl-3-pyridinemethanol; $C_8H_{12}N_2O_2$]. Stored in very limited quantities within the body, this water-soluble vitamin is readily available in major dietary sources including meat, poultry, fish, eggs, butter, and organ meats and in minor sources including potatoes, sweet potatoes, whole grains, legumes, nuts, tomatoes, and spinach. Vitamin B_6 is stable after exposure to heat and acids but easily destroyed by alkalis and light. All three forms of this vitamin are converted via enzymes to the biologically active form of vitamin B_6, pyridoxal 5′-phosphate. This active form is the coenzyme of a large group of specific enzymes that catalyze reactions of amino acid transfer, decarboxylation, and other metabolic transformations of individual amino acids. Additionally, this vitamin is converted in the body to coenzymes needed for the activity of enzymes involved in the metabolism of nucleic acids and lipids. It is also required for the conversion of the amino acid tryptophan to niacin, for glycogenolysis (the breakdown of glycogen to simple sugar glucose molecules), and for the formation of antibodies (crucial for immune system function) and hemoglobin (an oxygen-transporting molecule).

Biotin (hexahydro-2-oxo-1H-thienol[3,4-d]imidazole-4-pentanoic acid; $C_{10}H_{16}N_2O_3S$), a urea-derivative water-soluble vitamin containing sulfur, is stored in minute amounts particularly in the liver, kidneys, pancreas, brain, and adrenal glands, mainly bound to proteins or polypeptides. Stable when exposed to heat, light, and acids, biotin is available in dietary sources including liver, egg yolk, milk, yeast, legumes, cauliflower, mushrooms, and nuts. It is also synthesized by enteric bacteria within the gastrointestinal tract. Biotin plays an indispensable role as a coenzyme for enzymes involved in numerous naturally occurring carboxylation (–COOH), decarboxylation, and deamination reactions. It is also essential for reactions of the Kreb cycle (cellular aerobic respiration), for the formation of nucleic acid purine residues and nonessential amino acids, and for the use of amino acids for energy. Thus, biotin is involved in the utilization of carbon dioxide and in metabolic processes that lead to the formation of cellular energy.

Isolated from mammalian liver and bacterial cultures by E. L. Rickes and colleagues in 1948, vitamin B_{12}, or cyanocobalamin (5,6-dimethyl-benzimidazolyl cyanocobamide; $C_{63}H_{88}CoN_{14}O_{14}P$), refers to a variety of water-soluble compounds composed of a complex crystalline tetrapyr-rol ring structure (a derivative based on the basic skeleton ring structure of corrin [$C_{19}H_{22}N_4$]) and a cobalt (Co) ion in the center. This vitamin is stable when exposed to heat but inactivated by light and strongly acidic or alkaline solutions. Vitamin B_{12} is synthesized almost exclusively by mi-croorganisms (bacteria such as *Streptomyces griseus*) and is found princi-pally in the liver, where vitamin B_{12} stores are usually sufficient to provide body needs for three to six years. Converted by the body into its bioac-tive forms, methylcobalamin and cobamamide, this vitamin cannot be absorbed into the bloodstream from the gastrointestinal tract, or used by the body, until it is combined with a protein manufactured by the pari-etal cells of the stomach called intrinsic factor. Available in dietary sources including liver, meat, poultry, fish, eggs, and dairy foods (with the excep-tion of butter), this vitamin is not found in any plant foods, as plants have little or no vitamin B_{12} chemical cobalamin analogs. The bioactive forms of vitamin B_{12} are coenzymes necessary for processing carbohydrates, proteins, and fats in nearly all body cell types, but especially within the gastrointestinal tract, nervous system, and bone marrow. Within the ner-vous system, this vitamin is required for maintenance of the protective and insulating myelin sheath surrounding nerve cells. Within the bone marrow, it acts as a coenzyme in the synthesis and repair of DNA, thereby aiding in the development of erythrocytes (red blood cells). In-terestingly, a condition called pernicious anemia may result from an im-paired absorption (usually lack of sufficient intrinsic factor, rather than true vitamin deficiency in most cases) of vitamin B_{12}, whereby red blood cells do not divide, which causes the blood to have decreased oxygen-carrying capacity. This leads to clinical symptoms including weakness, weight loss, pallor, fever, and neurological impairment (e.g., numbness of extremities). Vitamin B_{12} is also essential for the synthesis of biochem-icals such as the amino acid methionine and the neurotransmitter build-ing block called choline.

Isolated by J. J. Pfiffner and colleagues in 1947, folic acid, or folacin and pteroylglutamic acid (N-[4-[[(2-amino-1,4-dihydro-4-oxo-6-pteri-dinyl)methyl]amino]benzoyl]-1-glutamic acid; $C_{19}H_{19}N_7O_6$), when in pure form, is a bright yellow crystalline water-soluble compound that is stable when exposed to heat but easily oxidized in acidic solutions or after exposure to light. Folic acid is a conjugated molecule consisting of a pteridine ring structure linked to *para*-aminobenzoic acid that forms pteroic acid. Stored mainly in the liver and kidney, this vitamin is synthesized by enteric bacteria located within the gastrointestinal tract and is readily available from dietary sources including liver, deep green vegetables, mushrooms, yeast, lean beef, eggs, veal, and whole grains. This vitamin is a coenzyme that is essential to virtually all biochemical reactions, acting as a carrier and transporter of single carbon atoms from one chemical to another. Such reactions include the biosynthesis of choline, of amino acids such as methionine, glycine, and serine, and of nucleic acid (e.g., DNA and RNA) building blocks such as purine. Specifically, folic acid interacts with vitamin B_{12} for the synthesis of DNA and acts in combination with vitamins B_{12} and C to assist in the breakdown of proteins and the formation of hemoglobin (a compound in red blood cells that may carry respiratory gases such as oxygen and carbon dioxide). Folic acid is therefore essential for proper embryological development during pregnancy, specifically neural tube development in the embryo, and for the proper formation of red blood cells within blood cell-generating (called hematopoietic) body tissues (e.g., bone marrow).

Vitamin C

In 1747, J. Lind discovered that a nutrient in citrus fruits assisted in preventing a disease called scurvy in British naval sailors during long sea voyages. The symptoms of scurvy include bleeding gums, anemia, easily bruised skin, degeneration of muscle and cartilage, osteoporosis, and weight loss. However, it was not until 1932 that A. Szent-Gyorgyi isolated a substance from the adrenal glands called hexuronic acid, and during the same period W. A. Waugh and C. King isolated a chemical from lemons and showed that is was identical to hexuronic acid. This acid, later termed vitamin C, was determined to be the vital compound in citrus fruits that prevented scurvy. The first vitamin to be artificially synthesized in a laboratory in 1935, vitamin C, also called ascorbic acid (l-ascorbic acid; l-3-ketothreohexuronic acid lactone; $C_6H_8O_6$), is a water-soluble vitamin that is a simple six-carbon crystalline compound derived from glucose via the uronic acid chemical pathway. The word "ascorbic" reflects the biological value of vitamin C in protecting against scurvy, as the words "ascorbic" and "scurvy" are derived from the Latin word for the symptoms of the scurvy disease (e.g., *scorbutus*). Rapidly destroyed by heat, light, and alkali compounds, a small amount of vitamin C is stored in body areas including the adrenal glands, retina of the eye, intestine, and pituitary gland located in the brain. Excess vitamin C is excreted in the urine when tissue storage sites are saturated. Dietary sources of this vitamin include fruits and vegetables, particularly citrus, cantaloupe, strawberries, tomatoes, fresh potatoes, and leafy green vegetables. The active form of vitamin C is ascorbic acid itself, and the main function is as a reducing agent in a number of different reactions. As an antioxidant, vitamin C has the potential to reduce cytochromes a and c of the respiratory chain as well as molecular oxygen, thereby preventing cell-damaging oxidation reactions. The most important reaction requiring vitamin C as a coenzyme is the hydroxylation (adding of –OH groups) of proline amino acid residues in collagen, a protein essential for the formation of nearly all connective tissues (e.g., dense connective tissue, bone, blood, cartilage, etc.) in the body. Vitamin C is therefore required for the maintenance of normal connective tissue as well as for wound healing and bone remodeling. It also is required for proper immune system function, for iron mineral absorption and use, and for various conversion reactions, including the conversion of the amino acid tryptophan to the neurotransmitter serotonin, of the amino acid tyrosine to the "fight-or-flight" hormone epinephrine, of the B complex vitamin folactin to its active form, and of the lipid cholesterol to bile salt acids essential for fat digestion within the intestine. Vitamin C is also believed to be involved in the process of steroid hormone synthesis (e.g., glucocorticoids, mineralocoticoids, androgen, estrogen) within the cortex of the adrenal glands located above each of the two human kidneys.

WART REMOVERS

Common cutaneous warts are hyperkeratotic papulonodules that most often appear on the hands, arms, and legs of an individual. The Latin word for wart is *verruca*, meaning "little hill" or "eminence." Of all the diseases that plague humankind, warts are thought by many to possess the highest number of folk remedy treatments. Throughout history, treatments for warts included application of plant extracts (e.g., willow bark), paring with a sharp penknife, burning with the ash of wine lees, use of corrosives (e.g., brimstone), and rubbing with pork fat, potatoes, green walnuts, broom straws, or intestines of black chickens. Ointments were often rubbed on warts and consisted of a variety of substances, such as dandelion juice, castor oil, and onion juice. Other types of treatments included rubbing warts with the blood of frogs or pigs, or with new pennies, or tying slugs to the wart. Most cures for warts, however, were based on the theory of transferring the warts to another person, animal, or object. While early eighteenth-century thought suggested that warts might be "congealed nutritious juices" that had seeped from damaged underlying nerve filaments up through the skin, it was not until the end of the nineteenth century that the infectious nature of warts was recognized. Soon afterward, in the early twentieth century, it was suggested that warts were caused by a virus, and by the 1950s, confirmation of this theory was possible with the visualization of virus particles using the electron microscope.

Warts are common, contagious, and usually benign proliferations of skin and mucosa caused by human papillomaviruses (HPV), which are double-stranded DNA viruses. The word papilloma is derived from *papilla*, meaning pustule or pimple, and the suffix *oma*, meaning tumor. HPV is widespread in human populations, and currently more than 150 types of HPV have been identified. Thus, infection of epidermal cells with HPV can manifest in benign cutaneous tumors as warts. Warts are filled with overgrown skin cells containing live HPV virus particles. Cutaneous (nongenital) warts are generally classified by their clinical features and morphology (e.g., common, flat, filiform) or location (e.g., plantar [undersurface of the foot]). Warts are usually spread by direct skin-to-skin inoculation of the virus from one person to another. Self-inoculation may also occur, especially with direct cutting of the wart, as any trauma that introduces breaks in the stratum corneum facilitates epidermal infection. Interestingly, the virus can resist desiccation, freezing, and prolonged storage outside of host cells. Thus, HPV may remain on and contaminate various surfaces (e.g., clothing, towels, floors, instruments) and subsequently infect scratched or broken skin via direct contact. While the incubation period for HPV ranges from one to six months, the latency period is suspected to last three or more years.

Because papillomaviruses are specialized for replication in external epithelia, infection is usually confined to the epithelium exposed to the external environment and does not result in systemic dissemination of the virus. While viral replication occurs in differentiated epithelial cells in the upper epidermal layer, viral particles are also often located in the deep epidermal basal layer. The epidermis becomes thickened and hyperkeratotic, and keratinocytes (keratin-containing cells) in the epidermal granular layer become vacuolated as a result of viral infection. The mechanisms by which virions penetrate the stratum corneum and infect viable keratinocytes is poorly understood, as there is a lack of practical in vitro culture systems available for these viruses to serve as study aids.

While the natural history of cutaneous HPV infections is for spontaneous regression within a few months or years, treatments (via medical personnel or the consumer) are frequently used to avoid potential viral spreading, wart enlargement, and individual social discomfort. Home remedies include the use of a pumice stone to remove the callus and/or warm water soaks (forty-five degrees centigrade) for thirty minutes daily for approximately six weeks. Most over-the-counter wart treatments function by chemically destroying the epidermis in which the virus is present and are recommended for the removal of only cutaneous common warts (with typical raised, rough "cauliflower-like" appearance) and plantar warts. These treatments include topical once- or twice-daily application of solutions or pastes containing keratolytic agents such as salicylic acid [2-hydroxybenzoic acid; $C_6H_4(OH)CO_2H$] and/or lactic acid as an active ingredient. By dissolving the intercellular cement substance, salicylic acid desquamates the horny virus-infected layer of the skin without affecting the structure of the viable epidermis. Treatment duration with this caustic agent depends on the size of the wart and the degree of skin discomfort tolerable to the consumer but generally requires up to twelve weeks. A flexible collodion (a liquid that dries to a long-acting film) or pad/disc-type product containing 15 to 40 percent salicylic acid is often applied directly to the wart lesion to lengthen chemical action time and allow enhanced chemical penetration into the skin epidermis. Such products can contain additional inactive ingredients such as solvents (e.g., acetone, alcohol, ether), fragrances (e.g., from camphor, menthol), emollients (e.g., polysorbate 80, propylene or polyethylene glycol), preservatives (e.g., quaternium-15, denatonium benzoate), thickeners (e.g., karaya gum base, ethylcellulose), and natural products historically thought to increase wart healing (e.g., castor oil, plant extracts derived from willow bark naturally containing salicylic acid). Therapeutic effects are usually enhanced by frequently removing dead surface keratin manually.

4

Baby Products

BABY OIL

Common baby oil is mineral oil derived from petroleum mixed with a small amount of fragrance. Baby oil has a number of uses, including prevention of scars, treatment for diaper and heat rash, sunburn relief, and treatment of canker sores. The purpose of using baby oil is to lock moisture into the skin. Typically, a small amount of baby oil is used after a bath or shower and increases the shine of the skin, making it appear healthier. Additionally, there are hundreds of formulations derived from plants, animals, and petroleum that are generically called "baby oil." Some animal oils, such as emu oil, has been used as baby oil. This natural formulation contains a large number of very diverse chemical compounds that have potentially useful properties. Emu oil is said to have antibacterial, anti-inflammatory, moisturizing, and skin-penetrating properties. Baby oil is a chemical substance that is so common it is often thought to be innocuous. Unfortunately, the hydrocarbons present in mineral oil can be very hazardous to a baby. If a child aspirates baby oil, the lungs become coated with the oil. This blocks oxygen from reaching the bloodstream by inhibiting proper respiration. The inhalation of baby oil can lead to serious cases of chemical pneumonia.

BABY WIPES

Because of the very thin, soft, and hydrated nature of baby skin, most types of chemical substances are easily permeable. In general, areas subject to a high concentration of moisture and soiling agents may provide a favorable environment for bacterial colonization. Areas of baby skin such as the buttocks, upper thighs, lower abdomen, pubic area, and groin that are subject to repetitive exposure to soiling agents, including urine and feces, require constant cleansing to avoid skin ailments (e.g., dermatitis).

Commercially available over-the-counter baby wet wipe products are manufactured to be minimally disruptive to the epidermal barrier and thus suitable for use on intact or compromised, irritated skin. While pre-moistened wipes generally differ in cleansing lotion formulation (emollients, preservative, pH, etc.) and fibrous composition, the wipes should be suitable for daily cleansing of the diapered area, even for infants with sensitive skin. Wipes should also be chemically formulated to avoid causing significant changes in the natural pH of pubic and buttock skin, as skin pH changes may exacerbate bacterial growth. In general, commercial baby wipes usually contain active cleansing ingredients such as purified water, potassium laureth phosphate, or chlorine dioxide. In addition, emollients and skin moisturizers such as propylene glycol, aloe barbadensis or aloe vera gel, polysorbate 20, cocamphodiacetate, sodium coco PG dimonium chloride phosphate, or glycerin may be added, along with emulsifiers such as cetyl hydroxyethylcellulose or PEG-75 lanolin. Vitamins (e.g., tocopheryl acetate [vitamin E acetate]), pH-buffering agents (e.g., tetrasodium EDTA, malic acid, citric acid), fragrance, and preservatives (e.g., sodium hydroxymethylglycinate, iodopropynyl butylcarbamate, DMDM hydantoin, methylparaben, propylparaben, 2-bromo-2-nitropropane-1,3-diol [alcohol], methylchlorisothiazolinone, methylisothiazolinone, quaternium 15, and potassium sorbate) may also be added. Some consumers have found the extensive use of alcohols, fragrances, and preservatives in commercial wipes to be drying and irritating to the delicate tissues of their children and choose to use homemade wiping cloths with a simple cleansing solution such as purified water, mild soap, and skin-lubricating mineral or natural herbal oil.

Cellulose fibers and an adhesive binder characterize the fibrous composition of most wipes. Often, special applications involve the depositing of cellulose fibers with the assistance of an electrostatic field to promote the properties of moisture absorbency and bulkiness with high absorption capacity. High-quality premoistened wipes are usually marketed in converted quarter-folded or flat-pack forms within a resealable plastic container.

DIAPER RASH TREATMENTS

Many factors contribute to the initiation of diaper rash, including excess moisture, rubbing and friction, skin contact with urine and feces, and/or allergic reaction to the diaper material or to creams, powder, or wipes. While true diaper rash (irritant diaper dermatitis) is most common in babies between the ages of four and fifteen months, incontinent adults risk developing this preventable skin problem as well, as the chronic use of diapers is the common factor in both population groups. In general, infant skin is much less of an effective barrier than that of children (over the age of three) and adults. Because of the thin, soft, and water-containing nature of baby skin, substances are more easily permeable. Areas subject to a high concentration of moisture and soil provide a favorable environment for bacterial growth. Thus, if these types of irritants remain in contact with infant skin over an extended period of time, a rash may develop. The occlusive nature of a diaper tends to inhibit the evaporation of moisture from the skin surface, eventually leading to skin decomposition and an increase in bacterial colonization. Some bacteria produce ammonia through the degradation of urinary urea, and ammonia can then be used as a nutritional substrate, resulting in the growth of even more bacteria. The added presence of feces may contribute urease, which also degrades urinary urea to ammonia. Ammonia will raise the pH of the skin, and this increase in alkalinity facilitates further bacterial growth. In addition, urine may enhance the irritant activity of chemicals by increasing the permeability of the skin and directly acting as an irritant. Diaper rash, characterized by reddened and warm skin, typically occurs in all areas in close contact with the diaper, including the buttocks, upper thighs, lower abdomen, and genitalia.

While prevention of diaper rash may be achieved by keeping the skin dry, preventing urine and feces from mixing together, and retaining an acidic pH on the skin, most cases of diaper rash are treated with products sold in toothpaste-like tubes or plastic jars obtained without prescription. Nearly all brands are formulated with skin protectants such as zinc oxide (ZnO; used in skin healing; antiseptic properties), petrolatum (a semi-solid mixture of hydrocarbons derived from petroleum), and/or dimethicone (silicone emollient). Other products added to diaper rash creams and ointments might include solvents (e.g., benzyl alcohol), opacifying agents (e.g., glyceryl oleate), lubricants (e.g., mineral oil, cod liver oil), emulsifiers (e.g., ozokerite, propylene glycol), humectants (e.g., sorbitol), preservatives (e.g., benzoic acid, borax, BHA, or methylparaben), fragrances, and additional skin protectants and/or wound-healing products, including allantoin, beeswax, silicone, calamine, kaolin, lanolin, and ceresin (earth wax). Specialty brands may also have added products such as vitamins (e.g., cholecalciferol [vitamin D], vitamins A and D [in cod liver oil], and vitamin E), talc (mineral; provides softness), topical starch (cornstarch), extracts of aloe vera (skin wound healing), Peruvian balsam (skin-healing stimulant; antiseptic), and/or bismuth subnitrate (forms a protective coating over inflamed skin areas). However, some of the above secondary ingredients have been known to elicit allergic reactions in some infants.

5

Cleaning Products

AMMONIA-BASED ALL-PURPOSE CLEANERS

Historically, ammonia (NH_3) was first synthesized from coal tar. However, these solutions were quite murky in appearance. In the early 1900s, before the onset of World War I, two German scientists, F. Haber and K. Bosch, developed the Haber-Bosch process, which involved the synthetic process of reacting nitrogen (N_2) gas and hydrogen (H_2) gas to form high quantities of pure ammonia gas. Natural gas (methane, CH_4) is first reacted with steam to produce carbon dioxide (CO_2) and hydrogen gas in a two-step process. Hydrogen gas and nitrogen gas are then reacted via the 1919 Nobel Prize-winning Haber-Bosch process to produce ammonia. These two scientists determined the conditions (extremely high temperatures and high pressures [steam]) and the catalysts (e.g., iron oxides and oxides of other common elements) necessary to produce ammonia gas. This colorless clear gas with a pungent odor is then easily liquefied for many uses. In fact, soap is often added to pure clear ammonia solutions to enhance consumer comfort compared with that of the originally marketed murky product.

Fresh household aqueous ammonia solutions range in concentrations of up to 10 percent actual ammonia. Such solutions are appropriate for use in loosening baked-on greasy soil or burned-on food particles. Diluted with water, ammonia solutions remove grease-based soils from glass surfaces, including mirrors and windows. Mixed with a detergent/surfactant, ammonia readily removes waxes from vinyl floor coverings (e.g., linoleum floors). This cleaner is not recommended for use on aluminum, asphalt tile, or woodwork, as it may lead to pitting, staining, and/or erosion of these materials. Ammonia-based solutions are excellent cleaners because they dissolve tough greasy stains without leaving a filmy residue. Both the ammonia and water evaporate after cleaning use. However, ammonia vapor is highly alkaline and caustic; therefore, it is potentially very irritating to the human respiratory system. In addition, it is extremely hazardous to mix ammonia with bleach. Potential noxious gases released from this reaction combination include chloramines (NH_2Cl, $NHCl_2$), hydrazine (NH_2NH_2), nitrogen trichloride (NCl_3), and hydrochloric acid (HCl), all of which are toxic.

DRAIN CLEANERS

Drain cleaners/openers are formulated to unclog kitchen sink and bathroom/lavatory drain traps. Most sinks are synthesized of steel coated with enamel, and the drain trap beneath is usually made of brass or polyvinyl chloride. Brass is an alloy consisting mostly of copper (Cu), with a key additional incorporation of zinc (Zn). Although copper is relatively unreactive toward both bases and acids, zinc reacts easily with acids and slowly with bases. When such drains become clogged, it is usually because of the accumulation of greasy/fatty soap scum, hair, and related unwanted products. Thus, the classic key active ingredient in drain cleaners is the caustic alkaline chemical sodium hydroxide ($NaOH$), also referred to as "lye," either in solid form with small chips of aluminum or as a concentrated liquid. When added to the product, the aluminum metal reacts with the sodium hydroxide solution to form hydrogen gas (H_2), which initiates a bubbling effect within the clog, thereby creating a stirring and agitation action.

Liquid and gel-based drain cleaners marketed for consumer household use can contain a combination of sodium hydroxide, sodium hypochlorite ($NaOCl$; bleach), sodium silicate (abrasive), and various detergents (surfactant cleaning action). In general, the sodium hydroxide reacts with the water in the pipe to generate a significant amount of heat, which melts away most of the greasy clog. The melted substances are thus broken down into simpler substances that can be rinsed away. A strong alkali will react with fat (e.g., triglycerides) to produce soap plus glycerine. Thus, the sodium hydroxide also reacts with some of the fatty grease, converting it to soap (i.e., saponification), which then is available to assist in cleaning and emulsifying the remainder of the grease within the pipe by a detergent action. If hair is the primary drain-clogging culprit (e.g., in shower drains), the bleaching agent included in the drain cleaner will degrade the keratin protein-based hair strands via oxidation, helping to unclog the drain.

Drains that simply possess slow water drainage, which are not truly blocked/plugged, can often be cleaned with slow-acting drain cleaners formulated with enzymes designed to degrade large insoluble organic (carbon-based) molecules into many smaller, water-soluble molecules. Examples of such enzymes include amylase (which breaks complex starches into glucose monosaccharide units), lipase (which breaks fats into glycerol and fatty acid units), protease (which breaks proteins into amino acid units), and cellulase (which breaks cellulose into glucose monosaccharide units).

GLASS CLEANERS

Glass cleaners are primarily liquids formulated to clean smooth glass-based areas such as windowpanes and mirrors. They loosen and dissolve oily soils located on glass, provide a shiny surface, and usually evaporate and dry quickly without leaving a streaking residue. One of the more common glass cleaner formulations contains isopropyl alcohol ($CH_3CH_2CH_2OH$ or $CH_3CHOHCH_3$) diluted with water. In some cases, ammonia (NH_3) or vinegar (acetic acid; CH_3COOH) is added for increased cleaning efficiency. Both ammonia and vinegar cut through greasy films, but alkaline ammonia vapors are highly irritating to the throat and entire respiratory system; thus, ammonia-based products should never be used for cleaning in a closed space. Ammonia-based glass cleaners do not leave a streaking residue, as both the ammonia and the water evaporate soon after application.

METAL/JEWELRY CLEANERS

Metal cleaners are formulated to remove various types of soils from metalware and leave behind a polished finish. Brass and copper (Cu) objects react slowly with environmental air to form compounds that result in a blackened and tarnished finish, which obscures the beauty of the metals. Tarnish, the end result of the oxidation of metal, is the principal type of soil found on most metalware. Metals primarily used for decorative jewelry (e.g., rings, bracelets, necklaces, etc.), including gold (Au), platinum (Pt), and silver (Ag), are also subject to chemicals (e.g., skin acids, sulfur-containing compounds) that create an oxidizing environment and subsequent tarnishing effect. In fact, most gold and silver jewelry items are almost always a mixture of metals. For example, most gold jewelry is alloyed with copper, zinc, nickel (Ni), or silver, and silver is frequently mixed with some copper and zinc, to strengthen the metals to withstand the wear and tear of usage. Platinum usually shows the least effects of oxidation, due to the purity and composition of this metal. Brass is simply a metal alloy composed mostly of copper, with a small inclusive fraction of zinc. As such, metal objects, which nearly all contain some amount of copper, may react with oxygen and sulfur compounds (e.g., hydrogen sulfide) to form CuO and CuS, respectively.

Although there are many techniques used to clean such objects, commercial products are often simple solutions of ammonia within a hydrogen-based solvent, with the additional inclusion of a very fine and mild abrasive called diatomaceous earth (DE). DE is nearly pure silica, in the form of SiO_2, with a very porous characteristic. DE consists of the skeletons of small aquatic unicellular algal organisms called diatoms, which have survived evolutionary processes for approximately 100 million years. Placed in the taxonomic family Bacillariophyceae, the cell walls of these creatures are made of silica. Because silica is more dense than seawater or freshwater, the presence of silica tends to cause diatoms to sink into the water depths. As such, DE is collected from the bottom of ancient lake beds and is currently mined and used for many commercial and industrial purposes. Thus, within metal cleaners, DE acts as an abrasive, and the alkaline ammonia dissolves any greasy residue on the metalware. In addition, the ammonia reacts with the CuO or CuS to form the soluble ammonia complex of copper, which is $Cu(NH_3)_4^{2+}$. The greasy tarnish residue can then be washed away with clean water and a damp cloth.

OVEN CLEANERS

Oven cleaners are formulated to remove burned-on and baked-on greasy soils and other food-based soils from the walls of ovens used for food preparation. Usually dispensed in aerosol form with thickeners and propellants, the primary active ingredient of oven cleaners is sodium hydroxide. These cleaners are formulated into a thick consistency to ensure that the product will cling to vertical surfaces within an oven. A strong alkali will react with fat (e.g., triglycerides) to produce soap plus glycerine. After oven cleaner application, the greasy deposits on the oven walls are converted to soaps upon reaction with sodium hydroxide. Thus, the sodium hydroxide also reacts with some of the fatty grease, converting it to soap (i.e., saponification), which then is available to assist in cleaning and emulsifying the remainder of the grease within the oven via a detergent action. The resulting mixture can be removed with a dampened sponge or cloth. The extremely high-temperature oven-cleaning cycle settings of many modern household ovens ensures the ease of greasy dirt removal once the saponification process is completed. However, it is recommended that the consumer wear protective rubber gloves while wiping off the mixture, as the highly alkaline sodium hydroxide is extremely caustic and potentially very damaging to fingernails and skin tissues.

SCOURING POWDERS

Scouring, or abrasive, cleansers are formulated to remove dense accumulations of soils commonly located in small household hard surface areas.

Most scouring powdered cleansers mainly consist of an insoluble abrasive powder (approximately 80 percent concentration), such as screened silica (SiO_2), feldspar, calcite, or limestone, with the size of the abrasive particles approximately 44 micrometers or smaller. The remainder of the product formulation often consists of calcium carbonate ($CaCO_3$) or similar alkaline salts (e.g., sodium carbonate), with an additional 2 percent grease-dissolving surfactants (e.g., anionic surfactants) and, in some cases, fragrance/perfume, color/dye and approximately 1 percent anhydrous chlorine bleach (e.g., sodium dichloro-*S*-triazinetrione dihydrate), which also acts as a disinfectant. Scrubbing with a small amount of water causes the abrasive ingredients to physically remove stains and deposits from hard surfaces such as porcelain tubs, sinks, cookware (e.g., pots and pans), bathroom fixtures, ceramic tiles, and outdoor grills. Organic-based stain material is then absorbed into the porous powder and rinsed away with the wash water.

Baking soda (sodium bicarbonate; $NaHCO_3$) is also a mild abrasive cleanser. This chemical absorbs food odors readily, making it appropriate for cleaning areas constantly in contact with foods, including certain types of countertops and the inside of the refrigerator.

TOILET BOWL CLEANERS

Toilet bowl cleaners are formulated to prevent or remove stains caused by hard water minerals (e.g., calcium [Ca], magnesium [Mg], etc.) and rust deposits resulting from iron (Fe) oxidation. These products allow for and contribute to the maintenance of a toilet bowl with pleasant odors, and some products contain active ingredients that disinfect. Most toilet bowls are manufactured from ceramic material with an applied tight glaze covering to prevent the soaking of water into the porous china-like materials. This glaze coating is relatively resistant to the decaying effects of acidic chemicals, so many toilet cleaning products contain a fair amount of acidic active ingredients. The buildup that forms within toilet bowls is mainly derived from calcium carbonate deposited from hard water minerals, along with discoloration attributable to iron compounds and the growth of fungal organisms. Acids readily dissolve calcium carbonate. The classic active ingredient for solid crystalline toilet bowl cleansing products is anhydrous granular sodium bisulfate ($NaHSO_4$). This chemical is mainly sulfuric acid for which one of the hydrogen ions has been neutralized by sodium hydroxide. The addition of sodium bisulfate to water within the bowl will yield a highly acidic pH of approximately 1, which provides an environment for the cleaning away of most hard water deposits, iron stains, and residual bowl-adhering fecal matter.

Many common liquid toilet bowl cleansers now contain 7 percent to 9 percent hydrochloric acid (HCl), citric acid (2-hydroxy-1,2,3-propanet-ricarboxylic acid; $C_6H_8O_7$), or some other acidic material. Hydrochloric acid is an especially good product for removing calcium carbonate-based hard water deposits and iron deposits (Fe_2O_3) because of the very high solubility of the chloride salts of calcium and iron in solution.

6

Lighting

CANDLES

Candles are an operationally simple device for providing heat and light by means of a controlled flame. A candle is made up of two parts, the fuel made of wax, and the wick made from absorbent string. The wax fuel source is typically paraffin wax that is a heavy hydrocarbon that comes from the refinement of crude oil. Paraffin wax is very flammable and, much like other oils from petroleum refinement, it must be very hot for burning to take place. Once the combustion of the wax has started, it can be very difficult to put out. A candle controls the burning process by allowing only a small amount of the wax to burn. The wick of the candle is made of an absorbent material that absorbs liquid wax and carries it to the end of the wick by capillary action. When a candle is lit, the heat of the flame melts the wax on and near the wick. The wick absorbs the liquid wax, and the heat of the flame vaporizes the wax. It is actually the vaporized paraffin wax being emitted from the wick that produces the flame observed. You may have noticed that you can relight a candle by touching a flame to the smoke leaving a candle just after it has been extinguished. This smoke is actually paraffin vapor that is still hot enough to exist in the gaseous form. Touching a lit match to this stream of vapor allows the flame to run down the vapor trail and relight the wick of the candle. It is also interesting to observe that the wick does not burn significantly. This is because the heat required to vaporize the wax actually cools the wick and protects it from burning. Paraffin wax consists of hydrocarbons with typically between eighteen and thirty-six carbons in their chain. This mixture of chain lengths is actually essential for the success of wax as a candle fuel. Shorter-chained hydrocarbons tend to have lower melting points and boiling points. The shorter-chain hydrocarbons tend to vaporize first and provide the heat necessary to vaporize the longer hydrocarbon chains. In this way, the mixture of chain lengths ensures an easy-lighting candle that burns for a long period of time.

FLUORESCENT LIGHTS

To understand how fluorescent lights work, it is helpful to remember that light results from the energy emitted when excited electrons return to their ground state. A fluorescent light contains a central element in a long sealed glass tube. This tube is typically filled with an inert gas, such as argon, and a small bit of mercury vapor. In addition, a phosphor powder is coated along the inside of the glass tube. When current is applied to the ends of the tube, electrons migrate through the gas from one end of the tube to the other. As the electrons move through the tube, some of them collide with gaseous mercury atoms. This collision promotes electrons in the mercury atom to a higher energy level. As these electrons revert back to their original energy level, they emit energy in the form of photons. The wavelength of the light emitted is a function of the energy difference between the ground state and the excited state. In a fluorescent tube, the mercury atoms release photons in the ultraviolet wavelength of the spectrum. This ultraviolet light is converted to visible light by the phosphors that coat the interior surface of the tube. When the photons emitted from the mercury collide with the phosphor coating, they transfer their energy, bumping the electrons of the phosphor to a higher energy level. The phosphor then releases photons as emitted light in the visible wavelengths. The phosphor is responsible for giving off the characteristic white light that we are familiar with. Varying compositions of phosphors can give different color fluorescent lights. As discussed later, incandescent bulbs waste a great deal of energy producing infrared light that is not seen by the human eye. Since fluorescent lamps yield a greater amount of visible light under the same circumstances, they are typically about five times more efficient than incandescent light bulbs. In contrast to traditional light sources, such as incandescent and halogen bulbs, fluorescent lights require a much higher voltage to operate. This is because of the greater resistance the electrons encounter moving through a gas instead of through a filament. It takes some time to heat the electrons in the tube to a temperature at which they will start to emit radiation in the form of photons. This is why fluorescent tubes sometimes take a second to light up when the current is applied. Transformers, which convert normal voltage to a much higher voltage, are used to apply higher-energy electrons to the light tube and ensure that the atoms emit photons of the proper wavelength. Fluorescent light tubes do contain a small amount of mercury vapor that can be toxic to humans. Therefore, great care should be exercised when cleaning up broken fluorescent light tubes.

GLOW STICKS

Light is a form of energy that can be emitted through a variety of processes. All of these processes rely on the same basic principle: energy excites an atom's electrons, and when these electrons return to their ground state they emit particles of light called photons. A light stick uses a chemical reaction instead of electricity to excite the atoms in a material. The chemical reaction typically involves several different steps. A typical commercial light stick holds a solution of hydrogen peroxide, phenyl oxylate ester, and a fluorescent dye. To produce light, the hydrogen peroxide oxidizes the phenyl oxylate ester to produce phenol and an unstable peroxyacid ester. This unstable ester decomposes, resulting in an additional phenol molecule and a cyclic peroxy compound. The peroxy compound then decomposes to form carbon dioxide. As a result of these decompositions, energy is released to the dye, where the electrons jump to a higher energy level. As they return to their ground state energy, they release energy in the form of photons that we see as visible light. The light stick is simply a housing for the two solutions involved in the reaction and prevents the chemical reaction from happening until the two solutions are allowed to mix with each other. The hydrogen peroxide is kept in a small glass vial inside the solution of dye and phenyl oxylate ester. When you bend the plastic stick, the glass vial is broken and the two solutions begin to react with each other. Many factors determine how long the reaction will last. As a result, light sticks may produce light from several minutes to as long as several hours.

HALOGEN LIGHTS

A halogen light is very similar in construction and operation to an incandescent light bulb. However, improvements in the design increase the amount of visible light produced and reduce the amount of infrared (nonvisible) light emitted. A halogen lamp uses a tungsten filament just like an incandescent bulb, but it is encased in a much smaller quartz envelope. This small envelope is so close to the filament that ordinary glass would melt. Inside this envelope, a gas from the halogen group, a group of reactive gases found in group 7 on the periodic table, is used to react with the heated tungsten metal. These gases, because of their high reactivity, combine with the tungsten metal vapor and redeposit them on the filament. This process of recycling the filament metal helps the filament last much longer than that in an ordinary incandescent light bulb. It is also possible to get the filament much hotter, producing more light. The proximity of the quartz envelope to the filament is why halogen lights become extremely hot compared with ordinary light bulbs.

INCANDESCENT LIGHTS

Light is a form of energy that can be released by an atom. Photons, the most basic unit of light, are made up of many small particle-like packets that have energy but no mass. Atoms release these photons when excited electrons revert back to their ground state energy. The excitement of these electrons, specifically how much energy is released, determines the wavelength of the emitted light and hence the color of the light. Incandescent light bulbs are very simple in construction and theory of operation. At the base of the light bulb, there are two metal contacts that connect to the ends of an electrical circuit. The metal contacts are attached to a thin metal filament, usually made of tungsten, and the globe of the bulb is filled with an inert gas to prevent the metal from burning in the presence of oxygen. When the bulb is connected to a power supply, an electric current flows from one contact to the other through the filament. As the electrons move through the filament, they continually bump into the atoms that make up the filament. This constant impact vibrates the atoms and heats the filament atoms to the point at which electrons may temporarily be boosted to a higher energy level. As these electrons return to their ground state, they release energy in the form of photons or light. The metal filament is heated to over 2,200 degrees centigrade, at which point the photons released are at a wavelength that humans can see, otherwise known as visible light. A great deal of energy is given off by incandescent lights that cannot be seen by humans. These wavelengths of infrared light are responsible for the heat that a light bulb emits. Approximately 10 percent of the light an incandescent bulb emits is visible; hence, incandescent bulbs are inefficient, wasting a majority of energy generating heat instead of light.

LIGHT-EMITTING DIODES

Light-emitting diodes, commonly called LEDs, function like tiny light bulbs and are found virtually everywhere. A diode is a simple semiconductor device that has a varying ability to conduct electrical current. In a LED, the conductor material is typically aluminum-gallium-arsenide (AlGaAs). In pure AlGaAs, all of the atoms bond perfectly, leaving no mobile electrons that could potentially conduct electricity. A "doped" semiconductor has additional atoms that either add mobile electrons or create holes where electrons can go. This doping process increases the conductivity of the material. Semiconductors doped with extra electrons are know as N-type materials, since they have extra negatively charged particles that tend to migrate toward positively charged areas. Semiconductors with holes are known as P-type materials, since they effectively have extra positively charged particles. Electrons jump from hole to hole, moving from a negatively charged area to a positively charged area. A diode is simply a section of N-type semiconductor bonded to a section of P-type material with electrodes on each end. When no voltage is present, the N-type material fills the holes from the P-type material at the junction between the two layers. In this arrangement, all of the holes are filled and the material behaves as an insulator. To get electrons moving, the N-type section is attached to the negative end of a circuit and the P-type section to the positive end of the circuit. The free electrons in the N-type section are repelled by the negative electrode and attracted to the positive electrode. The holes in the P-type material move in the opposite direction, migrating toward the negative end of the circuit. As such, this arrangement only conducts electricity in one direction. The interaction between electrons and holes has an interesting effect in that it produces light. As the electrons move across a diode, they can fall into empty holes in the P-type layer. As these electrons drop from their mobile state to a lower orbital, the electrons release energy in the form of photons. For example, the atoms in a standard silicon diode are arranged so that the electrons release a relatively small amount of energy, producing infrared light. These diodes are ideal for use in remote controls, among other products. The gap between the conduction band and the lower orbitals determines the frequency of the photons emitted; in other words, it controls the color of light produced. By adjusting the properties of the materials used for the P-type and N-type semiconductors, virtually any color of light can be produced from LEDs. The main advantage of LEDs is their efficiency compared with normal incandescent lights. These devices are tailored to produce a specific wavelength of visible light. As such, they produce very little heat because they do not produce infrared light unless they are intended to do so.

NEON LIGHTS

The technology behind how a neon light works is very different from that of normal incandescent lights. Electroluminescence, or the conversion of electricity directly into light, is the operating principle of neon lights. Neon lights are used in advertising signs and are made of long narrow glass tubes that are often bent in different shapes. These tubes can emit light in a variety of colors. The construction of a neon light is much like that of a fluorescent light tube. The glass tube is filled with a gas such as neon, argon, or krypton at low pressure. At both ends of the tube are metal electrodes. When a high voltage is applied to the electrodes, the gas ionizes, causing electrons to flow through the gas. These electrons excite the gas atoms and cause them to emit photons that we can see in the form of visible light. Neon gives off a characteristic red light, and other gases emit other colors of light. The main difference between a neon light and a fluorescent light is the lack of a phosphor coating. In a neon light, the visible light is produced directly from the excited gas and is visible to the human eye. Another example of electroluminescence is seen in Indiglo watches and alarm clocks. These operate in a very similar manner to neon lights. A high voltage is applied to a thin panel that is coated with a layer of a conductor and a layer of phosphor. When the voltage is applied, the phosphor emits visible light without emitting heat.

7

Common Household and Lawn Products

ALUMINUM CANS AND FOIL

In the 1800s, aluminum was believed to be so rare that it was considered more valuable than gold or silver and used only for jewelry. This is surprising, since aluminum is the third most abundant element found on earth. In the early 1900s, extraction procedures had improved greatly, making aluminum widely available at reasonable cost. Ideas for using aluminum in food storage began to evolve because of the favorable properties of this metal. Aluminum is lightweight, nontoxic, and easy to shape, it does not rust, and it can be easily recycled. These properties made aluminum an ideal material for packaging. In the 1950s, Coors Brewery partnered with a company called Aluminum International to develop an aluminum can for beer. This product was first marketed in 1958 and is the first commercial use of an aluminum beverage can. By 1964, a competing company, Reynolds Metals Co., became the first commercial supplier of twelve-ounce aluminum beverage cans that are still considered the standard today. Another development based upon the desirable properties of aluminum was thin foil. This aluminum foil was superior to papers and plastic because it protected contents from moisture and oxygen. Today, aluminum cans and foils are found with coatings of plastic to further protect the integrity of the stored product.

FERTILIZER

Plants require a number of different chemical elements to grow. The elements carbon, hydrogen, and oxygen are readily available from water and air and are required for plants to thrive. Other essential elements required by the plant are called macronutrients and include nitrogen, phosphorus, and potassium. These elements are basic building blocks for the amino acids that make up proteins and ultimately plant cells. These elements are the key to healthy plants. Molecules that make up the membranes of cells all contain phosphorus (phospholipids), nitrogen is important for the synthesis of amino acids, which all contain nitrogen, and potassium is essential to the metabolism of plant cells. If these macronutrients are absent, it severely limits the growth rate of the plant. The decay of dead plants in the soil is a source of these essential elements. Nature recycles the elements from dead plants to produce healthy growing plants. The purpose of fertilizers is to provide the elements that the plant needs to grow in easily available forms. Many fertilizers supply nitrogen, phosphorus, and potassium. The availability of these macronutrients seems to be the major limiting factor in plant growth. Carefully reading a bag of fertilizer, one will notice a series of three numbers on the bag. These numbers tell you the available percentage of nitrogen, phosphorus, and potassium found in the fertilizer. Although fertilizers mainly provide these macronutrients, a number of other elements, called micronutrients, are required in much lesser amounts to help plants grow. When looking at the periodic table, it is interesting to observe that only twenty-five of these elements are considered to be essential for plants and animals. Trace elements required for normal plant growth include boron, copper, iron, manganese, zinc, and molybdenum. Copper is important in the reproductive stage of the plants. Deficiencies in copper limit the yield and quality of the fruits and seeds, the products of reproduction humans typically consume. Manganese and molybdenum are essential for nitrogen metabolism and fixation. Iron is critical for photosynthesis and respiration. Zinc is essential for sugar regulation and enzymes that control plant growth. One danger of fertilizers is inherent to the available nitrogen source. Nitrogen in the form ammonium nitrate is the most widely used and can make up 10 to 40 percent by weight of a bag of fertilizer. Unfortunately, ammonium nitrate is a powerful explosive, and fertilizers can be misused with disastrous results, as in the Oklahoma City bombing.

FIRE EXTINGUISHER

Fire is a chemical reaction between the oxygen in the atmosphere and a fuel source. Fuels do not catch fire simply because they are surrounded by oxygen; the fuel must be heated for the combustion to take place. When a fuel is heated to its ignition temperature, the heat starts to decompose the fuel and release volatile gases. These gases formed from the decomposition of complex molecules react with oxygen to form water, carbon dioxide, and other products. The gases rise up through the air and make up the flame that emits heat and light. The heat produced by combustion keeps the fuel at the ignition temperature, so the fire continues to burn as long as fuel and oxygen are still present. The three essential elements to the fire are heat, oxygen, and fuel. Fire extinguishers are designed to remove one or more of these essential elements so the fire will cease this self-sustaining combustion. One of the best ways to remove heat is by dumping large amounts of water onto the fire. This cools the fuel below the ignition temperature as the water is converted to steam, interrupting the combustion cycle. It is also possible to remove the oxygen surrounding the fire by covering the fuel with a nonflammable material such as baking soda. This smothering effect separates the fuel from the atmospheric oxygen, again interrupting the combustion process.

Modern fire extinguishers operate on these two principles: removing heat or oxygen from the fire. A fire extinguisher is a metal cylinder that is filled with water or a smothering material with a mechanism to deliver the material from a safe distance. The extinguisher usually expels the water or smothering agent by pressure. A siphon tube leads from the bottom of the metal cylinder to the top of the extinguisher. The area above the active agent is pressurized with carbon dioxide. When the handle of the extinguisher is depressed, the carbon dioxide pressure forces the active agent through the siphon tube and out the nozzle of the extinguisher. If the extinguisher is aimed toward the fire, the water or smothering agent is delivered to cool or smother the fire.

Water is by far the most commonly used fire suppression material and one of the most effective. However, it is important that water not be used on fires involving electricity or flammable liquids. Most flammable liquids are less dense than water and will continue to burn while floating on top of the water. In these situations, the water serves only to spread the fire and will never effectively remove the heat as intended. Since water will also conduct electricity, there is the potential of electrocution should a water extinguisher be used on an electrical fire. Water extinguishers are intended to extinguish fires in which wood, plastic, or paper are the primary fuels and should not be used on electrical or flammable liquid fires.

Another popular extinguishing material is ordinary compressed carbon dioxide. Because carbon dioxide is heavier than air, it displaces the oxygen surrounding the fire, halting the combustion of the fuel. However, carbon dioxide extinguishers do little to cool the temperature of the fuel, and the carbon dioxide quickly dissipates. This can result in the fire reigniting. Ordinary baking soda (sodium bicarbonate), potassium bicarbonate, and ammonium phosphate are common smothering agents used in "dry chemical" fire extinguishers. These bicarbonates decompose and release carbon dioxide that displaces oxygen. Combined with the insulation from the dry material, it works by smothering the fire.

Most fire extinguishers contain small amounts of fire suppressant material and should only be used on relatively small fires. Some of these agents work by displacing oxygen from the atmosphere surrounding the fire. This can be hazardous to persons in the area, who may become asphyxiated because of the lack of breathable oxygen. The active agents in dry chemical extinguishers can be extremely hazardous if inhaled.

INSECTICIDES

Insecticides are used by farmers and homeowners to reduce the effect of various insects on crops, flowers, lawns, and ornamental plants. There are virtually hundreds of insecticides, many of which were developed as a consequence of early research on nerve gas weapons and which work to eliminate insects in a variety of ways. The way in which a particular pesticide kills an insect is called the mode of action and is a convenient way to classify the numerous available pesticides. Most traditional insecticides work on the nervous system. These chemicals are typically absorbed through the insect's skin and affect the nerve impulses of the insect, causing paralysis and brain death. Some pesticides act to slow the production of energy needed for the insect to survive and are typically used in fumigation; the insects die after a period of lethargy. Insects have an external skeleton called an exoskeleton, of which the protein chitin is a major component. Chemicals that inhibit the synthesis of this protein kill pests because they cannot shed their exoskeleton and grow a new one. These are very specific for certain pests and have been used with great success against fleas and termites. Unfortunately, pesticides affect humans just as much as the insects they are intended to kill. The body weight of humans is much higher than that of insects, so a much smaller dose is required to kill an insect than to kill a human. However, pesticides and chronic exposure can lead to very severe health problems. Many of these chemicals can be absorbed through the skin and are acutely toxic if inhaled. As such, appropriate precautions should be taken when using insecticides, such as wearing protective garments and always following the manufacturer's instructions.

WATER SOFTENERS

Most people are familiar with discoloration or stains that are a result of problem water in the home. Red and brown stains caused by iron, blue and green stains caused by copper, and white scales caused by magnesium and calcium are found in most homes with problem water. The process of softening water removes the minerals calcium and magnesium that are typically found in potable water. The removal of these elements is essential not only to make the water taste better but to prevent problems associated with the deposits of these minerals in the water pipes. If these minerals are allowed to build up in pipes, they reduce water flow and pipes can eventually become completely clogged. One solution is to distill the water or to use water filters to remove the minerals. This approach is impractical because of the high cost associated with filtering or distilling all of the water used in the household. A more cost-effective approach, water softening, removes magnesium and calcium by a replacement process. A water softener contains a bed of small plastic beads or a zeolite matrix, which is saturated with sodium chloride. As water flows through the softener, the calcium and magnesium are replaced by sodium ions. This process is called ion filtration. The undesirable calcium and magnesium remain in the softener and sodium ions are introduced to the water supply. Sodium does not cause precipitation problems, and the effects of hard water are eliminated.

8

Automotive and General Repair Products

AIR BAGS

Air bags have been mandatory in automobiles since 1998 and have decreased the risk of dying in an accident by nearly 30 percent. The technology behind air bags was first proposed during World War II to prevent injuries during a crash landing in an aircraft. Typically, air bags are located on the steering wheel or on the dashboard. They are intended to slow a passenger's speed to zero with little or no damage to the person. The air bag is made up of three major components. The bag is made of a thin nylon or equivalent fabric, which is folded compactly behind the dashboard or inside a small space in the center of the steering wheel. It typically contains a small amount of cornstarch or talcum powder to keep the nylon pliable during storage. The inflation system consists of a reaction between sodium azide and potassium nitrate to produce a large volume of nitrogen gas to inflate the bag. The sensor, which indicates the collision, is located in the bumpers of the automobile. The sensor is a switch that closes an electrical circuit when a collision above ten to fifteen miles per hour is detected. The electrical circuit triggers the chemical reaction in the inflation system and the air bag inflates in under one-twenty-fifth of a second. The reaction of the inflation system is so fast that the air bag is inflating at over 200 miles per hour! For this reason, it is important that passengers be seated at an appropriate distance from the air bag to avoid potential injuries from the air bag itself. Air bags can seriously harm children from the rapid inflation, and it is recommended that children under the age of one year not ride in the front seat of a vehicle equipped with air bags.

FUEL ADDITIVES

Since the mid 1920s, gasoline for consumer engines has contained the additive tetraethyllead, which improves fuel performance by preventing "knocking" in the cylinders of the engine. This knocking or preignition reduces engine efficiency, damages pistons, and reduces the power output of the internal combustion engine. Tetraethyllead, the most common fuel additive of the past century, is a toxic liquid that killed more than fifty chemical workers during its early development and manufacturing. Nevertheless, motor companies, oil companies, and the government authorized the manufacture and used of tetraethyllead in gasoline throughout the world.

In the late 1960s, new antipollution initiatives were enacted to reduce nitrogen oxides, carbon monoxide, and lead pollutants from automotive exhaust. Nitrogen oxides were responsible for the brown haze that hung over cities that can still be seen today. The advent of the catalytic converter, a small canister that contained heavy metal catalysts embedded on a ceramic support, helped oxidize carbon monoxide and reverse the reaction that produced nitrogen oxides. However, lead in the exhaust stream deactivated the catalysts in the catalytic converter. The only solution was to remove tetraethyllead from the gasoline.

Since the 1970s, tetraethyllead has not been added to gasoline, and oil refiners were pressed to increase the octane value of gasoline. One option involved externally oxygenating fuels by adding alcohols such as methanol, ethanol, and tertiary butyl alcohol as well as ether combinations such as methyl tertiary butyl ether (MTBE). The petroleum industry preferred MTBE to alcohol blends because it was somewhat easier to handle. Ethanol was seen as a viable solution but has long been resisted by the petroleum industry. Along with boosting octane, oxygenated fuels have lower hydrocarbon and carbon monoxide emissions and have been used to fight city smog since the late 1980s. The reformulated gasolines (RFGs) proposed by the petroleum industry reduce pollutants such as hydrocarbons, toxic aromatic compounds, and nitrogen oxides from the combustion of gasoline. RFGs reduced hydrocarbon emissions by at least 15 percent in major cities and reduced the cancer risk associated with gasolines with high benzene content.

Reformulated gasoline blends have several advantages over gasohol (gasoline-alcohol mixtures) and MTBE blends. RFG blends evaporate less readily because they have a lower vapor pressure, and they have lower benzene and sulfur contents. Gasoline blends, as well as other petroleum distillates, are an inhalation hazard to humans. Hydrocarbon distillates have been shown to cause cancer in laboratory animals. The products from the combustion of gasoline, specifically carbon monoxide and nitrogen oxides, have been linked to the dissipation of the protective ozone layer surrounding the earth and are responsible for a significant number of health problems in cities plagued with smog.

GASOLINE

Gasoline is a mixture of petroleum hydrocarbons (organic molecules composed of carbon and hydrogen) that in general contain chains of four to twelve carbons. It is primarily used as a motor vehicle fuel, although it is also used as a solvent in industry. The first cars were fueled by kerosene, but they knocked and were notoriously unreliable and needed constant repair. Scientists realized that the premature ignition or knocking was attributable to vapors present in the hot engine. Gasoline from the distillation of coal tar eventually replaced kerosene as an automotive fuel, and when combined with antiknock agents such as tetraethyllead became a standard fuel for the automotive industry in the late 1930s. Concerns about environmental pollution from automobiles burning leaded gasoline paved the way for the unleaded gasoline formulations of the 1970s, and lead was completely removed from gasoline in the early 1990s with the advent of the Clean Air Act.

When gasoline is burned in an internal combustion engine, carbon dioxide, water, and heat are produced. A gallon of gasoline contains approximately 13 million kilojoules of energy and is primarily made up of hydrocarbons with seven to eleven carbons. The gasoline vapor is ignited by the spark plug of the engine to drive the pistons, which transfer the energy to drive the automobile. Pollutants such as unburned gasoline hydrocarbons as well as nitrogen-containing compounds are removed in the catalytic converter. The familiar octane number that consumers see at the gasoline pump refers to gasoline's tendency to produce knocking in the engine. Isooctane is arbitrarily given a rating of 100 and n-heptane is given a rating of 0. When different hydrocarbons are blended or mixed with gasoline, the combustion in the cylinders of the engine is measured and compared with the octane scale. The higher the rating, the more efficient the combustion in the engine. Higher-octane gasolines are blended to produce very efficient combustion, which is why consumers pay more at the pump for high-octane gasoline.

Gasoline is extremely flammable in the liquid and vapor phases. It can accumulate a static charge through flow or agitation, potentially causing an explosion if the static is not dissipated. In addition to the physical hazards, gasoline is a carcinogen and causes central nervous system depression. The vapors are harmful if inhaled into the lungs. Coal tar also produces other fractions that include kerosene, diesel fuel, home heating oil, Vaseline, paraffin wax, and tar. These compounds all differ in the average length of their carbon chains, and their production depends on the fractional distillation of coal tar or crude oil.

MOTOR OIL

The distillation of crude oil or coal tar produces high-boiling components that are commonly used as lubricants such as motor oil. Solvent extraction and hydrogen refining are used to remove unwanted components and to increase the percentage of saturated hydrocarbons present. Modern lubricants must have good viscosity and temperature characteristics to meet the requirements of modern internal combustion engines. Corrosion and oxidation inhibitors are also added to increase the service life of the engines.

The discovery of petroleum oil's lubricant properties dates back to the 1850s, when it was discovered that oil was able to withstand high and low temperatures without losing its lubricant properties. In the late 1860s, a method of steam heat distillation of crude oil was used to obtain high-viscosity petroleum oil that was capable of withstanding the high heats associated with maintaining the lubrication of combustion engines of the era. Vacuum Oils Co., the forerunner of Mobile Corp., successfully marketed these petroleum lubricants to machinery owners to reduce wear and repair costs on expensive machinery used in millwork. These oils had extraordinary performance characteristics, such as the ability to reduce friction and wear, the ability to function dependably at temperature extremes, and the ability to withstand rigorous and lengthy engine operation without chemical breakdown. These properties would be invaluable in the new internal combustion engines of the time.

A machine's sliding or rolling surfaces must be separated to avoid friction and severe wear. This was accomplished by using lubricants with viscosities. Lubrication exists as long as a continuous thick film of oil separates the solid surfaces at all the points of wear, cooling and preserving metal parts from oxidation. In the internal combustion engine, oil must reliably lubricate bearings within the crankcase, pistons, and piston rings at temperatures at which fuel is being burned within the cylinder. Since oils change viscosity with respect to temperature, the ideal oil will protect moving parts equally at all temperatures. The principle of hydrodynamic lubrication, or coating the metal surfaces with a film of oil that will adhere to the surface, is essential to avoid piston wear in the engine. This thin coat of oil reduces friction and as a consequence the operating temperature of the engine, preserving the moving metal components in the engine.

Motor oils are hazardous to the environment and to humans having direct contact with them. Although motor oils can withstand greater temperatures than other petroleum products, such as gasoline, they are still highly flammable. Like most other petroleum distillates, motor oil is carcinogenic, mutagenic, and has adverse effects on the human reproductive system. Because of its higher boiling point, it is not a significant inhalation hazard and tends not to be absorbed through the skin readily. Prolonged exposure can lead to contact dermatitis. Motor oil that has been used should be discarded with care to avoid environmental contamination. Motor oils are less dense than water and are immiscible with water. Environmental contamination from oil is problematic in that there is no natural mechanism to break down the oil, so it accumulates in sedimentary deposits and eventually finds its way into the food chain, causing toxic effects. A typical barrel of crude oil ends up as gasoline for cars, kerosene for aircraft, diesel fuel for heavy vehicles, and lubricant oil for engines. A myriad of everyday products use products or by-products of the refining process.

RADIATOR FLUID

The active ingredient in radiator fluid is ethylene glycol, a two-carbon diol, which is used because of its low freezing point and high boiling point. A fluorescein dye is often added to readily identify antifreeze and make it less palatable to animals. Radiators are used with internal combustion engines to cool the engine block and pistons and maintain their specified operating temperatures. An engine's water pump circulates the coolant through the engine block and back to the radiator, where the heat is dissipated by passage through air-cooled coils. Initially, water was circulated to cool automobile engines. In areas with cold winters, the water would freeze and destroy the delicate thin-walled coils in the radiator, causing leaks that lead to overheating. Ethylene glycol has a freezing point lower than the frigid temperatures that caused water-cooled engine problems. Ethylene glycol has the added advantage that the boiling point is higher than that of water, allowing the fluid to operate over a wider range of temperatures than alternative substances. The temperature of the engine is controlled by the radiator fluid removing heat from the engine through the radiator, and the maximum coolant temperature is controlled with a high-temperature thermostat to control the flow rate of the liquid cooling the engine.

Ethylene glycol has a characteristically sweet smell that makes it irresistible to animals. The bright green or red fluorescein dye is added to warn humans and animals of the inherent danger associated with the ingestion of antifreeze. If antifreeze is swallowed, it causes central nervous system depression followed by respiratory and cardiac distress. If untreated, ingestion leads to cardiac failure, renal failure, and brain damage.

SUPERGLUE

Superglue is ethyl-2-cyanoacrylate, which when applied to surfaces bonds and polymerizes, forming a strong bond between materials. It was discovered accidentally by an employee of Kodak Research Labs who was trying to develop an optically clear plastic for gun sights during World War II. He found that everything that the acrylate monomers touched would stick together so well that it was difficult to get the pieces apart once the acrylate cured.

Superglue is an acrylic resin that bonds to most materials instantly, and the polymerization is triggered by the hydroxide ions in water. It undergoes a process called anionic polymerization, in which the cyanoacrylate monomers link together when they come into contact with water. The chains form a durable plastic mesh that continues to cross-link until the polymer strands can no longer move. These polymers bind the surfaces being glued together by millions of polymeric strands, giving the bond exceptional strength. It is primarily used in repair work on consumer items but has recently been used successfully by forensic examiners to develop latent fingerprints and to close surgical wounds without stitches.

The polymerization process of cyanoacrylates is so fast that it can react with the water present on skin, causing it to bond skin instantly. As such, it should be handled with extreme care, because a small amount rubbed in the eye can instantly bond eyelids shut. It causes irritation of the nose, throat, and lungs from the noxious vapors of the acrylate monomer. There are different commercially available chemical variants that have slightly different setting properties and viscosities depending upon the application.

SYNTHETIC MOTOR OIL

Synthetic motor oils are made of a synthesized hydrocarbon base oil of hydrogenated polydecene, decanoic acid esters, zinc alkyl dithiophosphate, and synthetic poly alpha olefins. Most synthetic oils also contain additives, detergents, and corrosion inhibitors as well as viscosity modifiers. It is believed that the first synthesized polymeric hydrocarbons were synthesized in 1877, yet it was not until 1929 that the commercial development of synthetic lubricants was undertaken. Because of the availability of commercial petroleum-based lubricants, these synthetic lubricants were ultimately unsuccessful. The advent of commercial jet travel spurred the development of the first commercially successful synthetic lubricant, Mobil 1, in 1975. This lubricant had superior resistance to thermal breakdown and lower friction properties than petroleum-based products.

Synthetic oils are used in engines to lubricate moving parts and protect them from wear and high temperatures. Detergents are added to synthetic formulations to reduce the deposits of contaminants on vital engine components, and corrosion inhibitors are added to reduce the effect of rust from the water produced in the combustion process.

Synthetic oils are somewhat safer than petroleum-based lubricants and do not cause any significant effects to the eyes, skin, or respiratory tract. Continuous exposure has been shown to cause cancer in laboratory animals. Brief or intermittent contact is not expected to pose any significant health problems in humans. Synthetic oils are related to petroleum-based lubricants with one significant difference. Whereas petroleum lubricants contain a variety of carbon chain lengths, synthetic lubricants contain specific compounds and polymers that help them adhere to the moving metal parts to combat friction. Other synthetic oils are used as hydraulic fluids, brake fluids, and antifreeze in automobiles.

WD-40

WD-40 is a proprietary formula composed of aliphatic petroleum distillates, petroleum base oil, carbon dioxide, and other nonhazardous ingredients. In 1953, the Rocket Chemical Company set out to create a line of rust prevention solvents and degreasers for use in the aerospace industry. On the fortieth attempt, they succeeded in formulating an effective water-displacing/lubricating formula, which they called WD-40. This product worked so well that it was used to protect the outer skin of the Atlas Missile from oxidation. It worked so well that employees of Rocket Chemical would sneak out cans of the formula for use in their own homes. The company produced a consumer version of the product in 1958, and since then people have used WD-40 on virtually everything.

WD-40 works by displacing moisture and forms a thin protective film on metal surfaces that protects the surface from oxidation and corrosion. It contains 60 to 70 percent aliphatic petroleum distillates, 15 to 25 percent petroleum base oil, and 2 to 3 percent carbon dioxide as a propellant. The volatile aliphatic hydrocarbons contained in the formula give WD-40 its characteristic odor and allow the formula to penetrate through rusted metal pieces. WD-40 has a much lower viscosity than standard light machine oils and is thought to contain tertiary alkyl amines, which have remarkably low surface tensions, providing low-temperature lubrication. Also important is the interfacial tension between WD-40 and water, because this is a measure of how fast the product will displace water out of the nooks and crannies of a rusty screw thread. This displacement is achieved with surfactants. Long-chain alcohols or quaternary ammonium salts may be some of the secret ingredients that help displace water and provide a temporary protective film to prevent corrosion.

WD-40 may cause headache or dizziness if inhaled and may cause drying or irritation of the skin upon contact. Eye contact and ingestion are especially dangerous and may lead to vomiting, severe irritation, and chemical pneumonitis if entering the lungs. As such, this product should only be used in well-ventilated areas with proper protective equipment such as gloves and eye protection. The petroleum base of WD-40 is essentially a narrow-boiling kerosene fraction and has a flash point of less than fifty degrees centigrade. It should never be used in the vicinity of open flames and should be considered extremely flammable.

WINDSHIELD WIPER FLUID

Windshield wiper fluid is composed of methanol, water, and a detergent to help remove dirt accumulated on automotive windshields. Methanol was first produced from the destructive distillation of wood—hence the common name "wood alcohol." Industrially, it is produced from natural gas (methane) but can also be produced from almost all organic materials, such as coal, wood, agricultural wastes, and garbage. High-temperature decomposition of these materials results in carbon dioxide, carbon monoxide, and hydrogen gases. These gases are catalytically converted to methanol, which is an important starting material for a number of industrial chemicals, such as formaldehyde plastics (Bakelite), antifreeze, rocket fuel, and polyester plastics.

The freezing point of methanol is –97.8 degrees centigrade; therefore, it remains liquid even at frigid winter temperatures. Mixing methanol with ice has the effect of depressing the freezing point of water; a mixture of water and methanol has a freezing point less than that of pure water alone. Thus, it helps to remove accumulated snow and ice on the windshield and prevents the windshield from becoming covered with ice. Methanol is highly toxic. Inhalation of methanol can lead to abdominal cramps, weakness, dizziness, and confusion. Ingestion of even small amounts of methanol can lead to blindness and death. Because of the inherent toxicity of methanol, several alternatives have been used for keeping windshields clear of ice and dirt. Ethanol-based washer fluids have been used successfully without the toxicity problems associated with methanol-based products.

WOOD GLUE

Wood glue, or polyvinyl acetate (PVA) glues, are the most common adhesives on the market. Glues come in a variety of formulas, all only slightly different, and specific to what they are designed to glue. In general, PVA glues are designed to work on porous materials and clean up easily because they are water based. They need pressure and air circulation to adhere to surfaces and maintain the strength of the bond. These glues are relatively safe consumer products because they do not emit any harmful fumes and are not hazardous to touch. PVA glues are only toxic if ingested.

9

Common Materials and Office Supply Products

ALKALINE BATTERIES

Alkaline batteries are common in electronic devices used in everyday life; however, the way in which they produce electricity is little understood. They are typically constructed with a high-surface-area zinc anode (the positive end) and a manganese dioxide cathode (the negative end). Inside the battery case is a potassium hydroxide electrolyte that is gelled with a cellulose derivative. This electrolyte allows electrons to flow from the anode to the cathode. When the circuit is completed, this flow of electrons powers the electronic device. Although there are many different types of batteries, the construction is typically the same. What varies from battery to battery is the composition of the anode, cathode, and electrolyte. Alkaline batteries were derived from the older "dry cell"-type batteries and were put into commercial production in the early 1950s. A battery is nothing more than a contained chemical reaction that creates an electron potential gradient. The zinc anode is oxidized to form zinc hydroxide-releasing electrons. The manganese dioxide at the cathode is reduced by electrons from the zinc, which are carried to the cathode through the electrolyte solution. Although this is the most common type of alkaline battery in use, some alkaline batteries use mercury and silver in their construction. These metals, especially mercury, can be highly toxic and can be problematic to dispose of properly. Batteries may rupture when exposed to excess heat that accelerates the chemical reaction. This may release harmful flammable or corrosive materials that can result in serious chemical burns and skin irritation.

CORRECTION FLUID

The invention of White Out and other correction fluids was spawned by the need of Edwin Johanknecht for a correction fluid that worked on photostatic paper. In 1971, Johanknecht and his partners incorporated their new business, White-Out Products, Inc., providing correction fluid to users all over the world. Correction fluids usually contain a proprietary mixture of solvents and whiteners that act much like a thin fast-drying paint. Typically, methyl chloroform (1,1,1-trichloroethane) is used as a fast-drying solvent in which pigments such as titanium dioxide and other colorants are suspended. Correction fluid is brushed across the paper to cover mistakes and once dry can be written or typed over. Although the product is marketed for consumer use, there are a number of hazards associated with it. Inhalation of large amounts of solvent vapor may cause unconsciousness and may irritate the mucus membranes of susceptible individuals. It can also have a long-term effect on the central nervous system and cause abnormal heartbeats in humans. The dried correction fluid can be especially hazardous if burned, releasing small amounts of phosgene, hydrogen chloride, and chlorine. The formula will react with strong bases and oxidizing agents as well as reactive metals. For these reasons, it is important that correction fluid be used according to the manufacturer's instructions.

PENCILS

The origins of the modern pencil date back to the mid 1500s, when graphite was discovered in the Seathwaite Valley in England. Soon thereafter, graphite was used as a writing instrument. The real breakthrough came when French chemist Nicholas Conte discovered a process of firing a mixture of clay and graphite in varying proportions to create writing instruments of different hardnesses. These kiln-fired mixtures were encased in wood, and the pencil was born. In the United States, William Monroe manufactured wooden pencils in the early 1800s. Mass production of pencils in the United States began during the Civil War, when machinery for the production of pencils began to make handmade pencils obsolete. The mass production of pencils led to other related advances, such as the rubber eraser, first patented by Hyman Lipman in Philadelphia, and the mechanical pencil sharpener. Pencils do not contain lead, as is commonly thought. Instead, graphite, a form of carbon, is used along with various binders to modify the hardness of the pencil.

PENS

The history of the ink pen dates back many centuries, to the days of writing with a quill and ink with a charcoal base. Modern pens are much different than the reed, quill, or fountain pen of centuries ago. These older designs had several significant drawbacks, such as the uneven flow of ink, the slow drying of the ink documents, and the laborious cleanup associated with keeping the pens in good writing condition. The first improvement was the switch to faster-drying ink. Quick-drying ink was already used to print newspapers, but it was not until a Hungarian journalist, Laszlo Biro, developed a pen that used fast-drying ink that pens became practical for consumers. Biro developed a pen that used a tiny metal ball rotating inside a metal tube to keep the ink in the pen from drying out and let the ink flow out at a controlled rate. These "ballpoint pens" were patented by Biro in 1943, with the patent eventually being acquired by the British government so that the pens could be used by Royal Air Force crews. Milton Reynolds produced the first commercial ballpoint pen in the United States in 1945, selling thousands of pens until the more inexpensive BIC ballpoint pens were introduced on the American market in the mid 1950s. The development of new inks for use in these pens was a laborious trial-and-error process. Inks using linseed, rosin, or wood oil are darkened with dyes to produce different colors. Carbon gives plant oil ink a black appearance, chromium produces green and orange colors, cadmium colors the ink red, and iron in the form of ferrous sulfate gives the widely used blue ink color.

POLARIZED SUNGLASSES

How light waves interact with matter has been studied at great length throughout centuries. Sunlight and other light sources transmit light as waves at perpendicular angles to the direction of travel. In 1669, a Danish mathematician, Erasmus Bartholinus, performed the first comprehensive experiments concerning the interaction of light with transparent calcite crystals, which is considered the first account of polarized light waves. In 1932, Edwin Land developed a synthetic film of quinine sulfate and iodine embedded in nitrocellulose and trademarked the name Polaroid. This material had a very interesting property. It allowed only light waves that were cohesive, or traveling in the same direction, to pass through the lens. In practice, this reduced the ambient light reaching the eye by more than 50 percent, making the material ideal for sunglasses. It also significantly reduced sun glare from objects by allowing only a fraction of the glaring light to pass through the lenses. Today, polarized sunglasses are made by stretching an amorphous polymer of polyvinyl alcohol that aligns the molecules and sandwiching this layer between sheets of transparent material such as glass. These sunglasses are used in a variety of activities, such as driving, skiing, boating, and playing sports. Polarized sunglasses help reduce glare and filter out a significant amount of light without distorting images.

RECHARGEABLE BATTERIES

Many consumers are familiar with rechargeable batteries that are common in many everyday devices, such as cellular phones, laptop computers, and digital cameras. These batteries are constructed very much like a normal dry cell or alkaline battery. Typically, they contain nickel, cobalt, zinc, or cadmium and an electrolyte of potassium, sodium, or lithium hydroxide. The most common rechargeable batteries are based on a nickel/cadmium design with a cadmium anode and a nickel oxide cathode. These batteries operate very much like other batteries with one major difference. When the chemical reaction has been completed and no electron gradient exists between the anode and the cathode, these batteries can be recharged by an external current that reverses the flow of electrons from the cathode to the anode. This reestablishes the electron gradient between the two ends of the battery and allows the battery to be reused after charging is complete. This process can be repeated as many as 1,000 times before the batteries lose their ability to be recharged. Rechargeable batteries are not as new as you might think. The most common rechargeable battery is the lead-acid battery, which has been found in automobiles for the better part of a century. Rechargeable batteries undergo a reversible reaction, providing electrons to power devices and being recharged by reversing the chemical reaction to pump electrons back into the battery. Rechargeable batteries may leak when recharged, and the contents are harmful if swallowed and cause burns from the caustic electrolyte materials. As such, it is important to be sure that batteries are never connected improperly, short circuited, or exposed to high temperatures.

RECYCLED PAPER

Recycled paper is paper that contains fiber from waste paper. The United States has more than 500 paper mills that currently use at least some recycled materials in production. Recycled paper is made by collecting large amounts of used paper and washing the paper to remove any ink that may be present. The paper is then literally beaten to a pulp by machines and bleached, most commonly with chlorine bleaches, to make the paper white. Once a uniform white pulp is obtained, the paper goes to a forming section, followed by a series of presses that remove the water from the pulp. The wet paper moves on to a drying stage, where the moisture content is reduced and the thickness is reduced by rolling through a number of steel rollers. It is estimated that for one ton of paper the recycling process saves over 7,000 gallons of water, 500 gallons of oil, and 600 pounds of air pollution! The practice of collecting paper for recycling is common today. The majority of recovered paper comes from businesses and industry, where the recovery rate approaches nearly 70 percent. Paper seems to be one of the most benign items with regard to human health. In the earlier years of paper recycling, a bleaching by-product called dioxin was found throughout the pulp and paper manufacturing process. Dioxin is one of the most toxic human-made chemicals known. Pulp and paper mills using chlorine-based methods of whitening create dioxin when the chlorine reacts with chemicals present in the wood and other fibers. It is interesting that chlorinated toilet paper contains the highest amounts of these harmful chemicals.

RECYCLED PLASTIC

Plastics are well known for their desirable properties of being lightweight, strong, and formed into a variety of shapes. In fact, modern plastics might just be too durable. Millions of tons of plastics are buried in landfills and will never be broken down by nature. To address the environmental impact of this problem, a vigorous recycling program has been enacted to reuse plastics and minimize their effect on the environment. Plastics recycling has been a great success for polyethylene terephthalate, which is used in soda bottles, and high-density polyethylene, which is used in numerous applications from plumbing to computers. One of the biggest problems of plastics recycling is the great diversity of plastics that are collected. Certain polymers do not mix, bond, or adhere well to one another. A recycled product derived from mixed plastics can have inferior strength and durability. In cases in which plastics cannot be separated into recyclable material, they may be used as fuels in incinerators because of their large heat value. To ensure success in the recycling processes, the source materials must be relatively clean, homogeneous, and free of contamination. In the recycling process, homologous plastics are reduced in size to very small pieces that are heated, cleansed of impurities, and formed into small pellets. These small pellets are sold to manufacturers and can be reincorporated into a variety of products. Polystyrene coffee cups are recycled and commonly used as inexpensive stuffing for toys or beanbag chairs. Polyethylene soda bottle material may be used to make trash cans, flower pots, traffic cones, and plastic lumber. Mixed plastics that are not incinerated are used to produce extruded plastic lumber, shipping pallets, fencing, and park benches.

RUBBER CEMENT

Rubber cement is commonly used as an adhesive that serves to bond two objects together with a degree of flexibility. In the early 1800s, a viscous mixture of natural rubber dispersed in benzene or gasoline was used to cement seams of garments to ensure that they would be watertight. This mixture was later used to completely waterproof wool and cotton garments. The ability of rubber cement to form a durable and flexible bond is the main reason it is chosen as an office adhesive product. Little has changed in the formulation of rubber cement from the 1800s. Because of the flammability and toxicity of benzene and gasoline, these solvents are no longer found in modern rubber cement formulations. Today, rubber cement is a mixture of natural or synthetic rubber dispersed in isopropyl alcohol and heptane. These solvents, although safer, should not be considered innocuous. Isopropanol may cause reproductive disorders, and heptane is extremely flammable. For these reasons, rubber cement should be used in a ventilated area and away from open flames. Inhalation may cause irritation to the nose, throat, and lungs. Ingestion is equally hazardous, causing burns to the esophagus and gastrointestinal tract.

SPRAY PAINT

The idea of delivering a controlled spray of liquid was developed in the early 1930s by Eric Rotheim. This delivery system is used for a number of applications in industry and in homes. The design of the spray can is very simple. The metal can contains a propellant, paint, and a nozzle for delivering the spray. You may have noticed that the bottom of a spray can has a concave shape. This is done to strengthen the can and allow pressure within the can to be distributed to the walls of the can. The propellant is a fluid or gas stored under pressure and expels the paint out of the can. The propellant is generally a substance that is a gas at room temperature and pressure. Under pressure, the propellant is liquefied and will remain liquid even though it is well above its boiling point. The other liquid, in this case liquid paint, is also stored in the sealed metal can. In the metal can container, a long plastic tube runs from the valve at the top of the can all the way to the bottom of the can. The curved shape of the bottom of the can ensures that no paint is wasted in the bottom of the can. The nozzle has a very fine opening that serves to break the liquid paint into small droplets, ensuring controlled coverage of the paint. When the valve on the nozzle is opened, the pressure within the can is reduced, allowing the liquefied propellant to form a gaseous layer at the top of the can. The pressure exerted by this gas layer forces the paint at the bottom of the can to travel up the plastic tube connected to the nozzle, where it is released as a fine spray. You may have noticed that spray cans instruct you to shake the can for several minutes before painting. A small plastic or metal ball inside the can mixes up the propellant and paint to ensure a uniform consistency of paint as it is expelled from the nozzle. Until the 1980s, most manufacturers used chlorofluorocarbons (CFCs) as propellants in spray cans. It has been concluded that the use of CFCs is harmful to the ozone layer, and CFC-based propellants were phased out. Today, the most popular alternative propellant is liquefied petroleum gas, which does not harm the ozone layer but must be used with care because of its extreme flammability.

TINTED WINDOWS

As energy consumption continues to grow in the United States, companies are looking to make homes more energy efficient. It is estimated that 2 percent of all energy consumed can be attributed to the inefficiency of household windows. Improving window technology could save billions of dollars spent on heating, lighting, and cooling costs for consumers nationwide. The concept of adjustable darkness windows in households began with the development of photochromic materials. Photochromic materials darken upon exposure to light and block ultraviolet radiation from the sun, reducing the strain on your air conditioner in the summer. These materials have been used for some time in consumer eyewear but have not made the transition to window coatings. The main reason is that such windows respond to light only and are not manually controlled. During the winter, photochromic windows would also exclude desirable warming ultraviolet radiation. An adjustable window film that could be darkened or lightened depending on conditions would be ideal to reduce energy consumption regardless of the season. Liquid crystalline displays (LCDs), which are found in watches, computers, and a variety of other electronic devices, work by aligning their crystals to allow light to pass through. In the absence of an electrical current, the liquid crystals are randomly aligned and appear as dark spots that form the numbers and letters that we see on an LCD display. LCD-based windows are already found in homes and offices around the world. With a flick of the switch, they turn from transparent to dark. These materials are used primarily in user-controllable privacy windows. The drawback to LCD-based devices is the lack of control over the darkness of the window. This problem is addressed in a new technology called a suspended particle device, which allows a window to change from dark to clear by applying an external electrical current. These devices are a sandwich of two glass panes coated with conductive material that hold a liquid suspension of particles. The particles align themselves when an external current is applied to the conductive glass coating, allowing light to pass through the glass. When the current is removed, the suspended particles move back to a random orientation to block light and darken the glass. Varying the applied voltage produces different shades of darkness for the glass.

Appendix
Laboratory Exercises

DENSITY DETERMINATION
Procedure:
DENSITY OF WATER

1. Weigh a dry, clean flask.
2. Pipet 10 ml of water into the preweighed flask.
3. Reweigh the flask with the added water.
4. Calculate the difference between the two weights (mass of water).
5. Calculate the density of water.

mass of flask + water   _____ g

mass of flask   _____ g

mass of water   _____ g

volume of water   _____ ml

Show the calculation for the density of water.

DENSITY OF VEGETABLE OIL

1. Place some vegetable oil and water together in a test tube.
2. Cover the test tube with Parafilm and shake (allow it to settle).

What layer is on the top?

Is oil more or less dense than water?

---

1. Weigh a dry, clean flask.
2. Pipet 10 ml of oil into the preweighed flask.
3. Reweigh the flask with the added oil.
4. Calculate the difference between the two weights (mass of oil).
5. Calculate the density of oil.

mass of flask + oil	_____ g
mass of flask	_____ g
mass of oil	_____ g
volume of oil	_____ ml

Show the calculation for the density of vegetable oil.

Compare your results with the density of water. Which is more dense, water or oil?

DENSITY OF ICE

Work quickly for this experiment—ice melts!!!

Why does ice float?

---

1. Place about 20 ml of water into a graduated cylinder.
2. Measure and record the initial volume.
3. Weigh ice cubes. (Make sure the ice cubes or crushed ice will fit into the graduated cylinder.)
4. Add the ice to the cylinder quickly and take another volume reading.
5. Calculate the difference in the two volume readings. This allows you to determine the volume of the solid by liquid displacement.

mass of ice   _____ g

final volume cylinder reading   _____ ml

initial volume cylinder reading   _____ ml

volume of ice   _____ ml

Calculate the density of ice.

_____

_____

Is the ice really less dense than water? Did the ice float or sink?

_____

_____

Practical Application

A piece of glass is found at a crime scene by police detectives that does not match any of the other glass or glasslike material present at the scene. The initial investigation by the police has also identified two possible suspects. A broken wine glass was found at the apartment of suspect 1 and was collected and marked into evidence. At the home of suspect 2, investigators found fragments from a glass coffee table top that the suspect said broke a month ago.

At the forensic laboratory, you have been assigned to determine if either piece of evidence can be linked to the material discovered at the crime scene.

1. Design an experiment based on the knowledge gained in this laboratory exercise to help solve this crime.

Give your procedure.

_____

_____

_____

2. Conduct your experiment.

_____

_____

_____

What can you conclude from your experimental findings? Explain.

_____

_____

_____

EVALUATION OF AN ENDOTHERMIC REACTION

Procedure:

1. Prepare 25-ml saturated solutions of citric acid and sodium bicarbonate (baking soda) by dissolving an excess of each in 25 ml of water and decanting the clear supernatant liquid.
2. Pour the saturated aqueous sodium bicarbonate solution into a polystyrene coffee cup. Use a thermometer to record the initial temperature of the solution. Rinse the thermometer with water and record the initial temperature of the citric acid solution.
3. Pour the citric acid solution into the coffee cup containing the sodium bicarbonate solution. Swirl the solution and monitor the temperature with a thermometer.
4. Record your observations.

GROWING CRYSTALS WITH SEED CRYSTALS

Procedure:

MAKING SEED CRYSTALS

1. Dissolve 10 g of alum (potassium aluminum sulfate) in 75 ml of distilled water.
2. Heat the solution to 60°C with stirring until the solution is clear. If it is not clear when hot, filter or decant the supernatant liquid and use it for the next step.
3. Cover the solution with aluminum foil or a watch glass and let the solution cool undisturbed for several days to a week.
4. After a period of time, crystals will precipitate out of the solution and can be removed carefully with a set of forceps.

GROWING HUGE CRYSTALS

1. Dissolve 10 g of alum in 75 ml of distilled water as above.
2. Let the clear solution cool to room temperature and add the seed crystal from the previous experiment.
3. Cover the solution and let it sit for a week to several weeks.

INFLUENCE OF SALT ON THE FREEZING POINT OF WATER

Procedure:

1. Weigh a clean, dry 250-ml beaker. Record the weight.
2. Put three ice cubes in the preweighed beaker.
3. Weigh and record the weight of the beaker and the ice.
4. Add 100 ml of water to the beaker.
5. Using the thermometer, measure and record the temperature of the ice solution.
6. Using a stirring rod (NOT the thermometer), stir the solution gently.
7. Record the temperature of the solution after 1 minute, 2 minutes, 5 minutes, and 10 minutes.
8. After 10 minutes, pour the water off the ice carefully (DO NOT POUR OFF ICE) and quickly weigh the beaker and ice.
9. Prepare the ice beaker as before using the same amount of ice except also add 1.7 g of NaCl (0.029 mol) to the new ice water mixture.
10. Repeat steps 5 to 8. Record your results.
11. Repeat steps 9 and 10 for 5.0 g (0.086 mol), 8.3 g (0.142 mol), and 11.7 g (0.200 mol) of NaCl. [Note: the molar mass of NaCl = 58.44 g/mol.] Make sure that the temperature at the beginning of each run is the original temperature and that the ice water mixture is prepared new.
12. Repeat steps 9 and 10 for 3.218 g (0.029 mol) of $CaCl_2$. [Note: the molar mass of $CaCl_2$ = 110.98 g/mol.] Make sure that the temperature at the beginning of each run is the original temperature and that the ice water mixture is prepared new.

Questions:

Which amount of NaCl caused the most ice to melt?

_____

Which melts ice faster, NaCl or $CaCl_2$?

_____

Results:

	Salt added					
	0 mol NaCl	0.029 mol NaCl	0.086 mol NaCl	0.142 mol NaCl	0.200 mol NaCl	0.29 mol CaCl$_2$
Original temperature						
Temperature at 1 minute						
Temperature at 2 minutes						
Temperature at 5 minutes						
Temperature at 10 minutes						
Difference in temperature						
Initial beaker + ice weight						
Beaker weight						
Initial ice weight						
Final beaker + ice weight						
Beaker weight						
Final ice weight						
Percentage of ice melted = {(initial − final)/ initial} × 100						

MAKING SLIME

Procedure:

1. Add 3 g of borax to 50 ml of water and stir until completely dissolved.
2. Add 25 ml of white glue to 25 ml of water and mix thoroughly.
3. Add the 50 ml of borax solution from step 1 to the 50 ml of glue solution from step 2 in a Ziploc bag.
4. Seal the Ziploc bag and knead the mixture.
5. Food coloring can be added to color the slime. Keep the slime in a sealed bag in the refrigerator when not in use.

MAKING YOUR PENNIES MORE VALUABLE: TURNING COPPER INTO SILVER AND GOLD

Procedure:

1. Obtain several new pennies. The cleaner and shinier the penny is, the better the experiment will work. Dirty pennies can be scrubbed with a test tube brush or soaked in 1 M acetic acid to remove surface oxidation.
2. In a small beaker or jar, add 30 ml of 1.0 M $ZnCl_2$ and 1.5 g of zinc metal.
3. Place your pennies in the beaker in contact with the zinc metal.
4. A gentle warming of the solution on a hot plate is necessary for the pennies to take on a silver color. The pennies can be turned over and the warming continued until a uniform silver color is achieved.
5. Remove the pennies and rinse them with distilled water. You will notice that you have turned your old copper penny into a silver penny. By carefully heating the penny in a Bunsen burner or on a hot plate, the penny will become a beautiful gold color. The mixture of copper and zinc that you have made is actually a brass alloy and not gold!

PAPER CHROMATOGRAPHY OF M&M COLORS

Procedure:

1. Prepare a 1.0 M solution of NaCl (aqueous) by dissolving 58.0 g of NaCl in 1.0 liter of distilled water.
2. Place each of several M&M colors in a few drops of 1 M NaCl to dissolve the candy coating.
3. Cut a piece of filter or chromatography paper into a strip approximately 2 inches wide by 4 inches long. Draw a light pencil line across the bottom of the filter paper approximately half an inch from the end of the paper.
4. Using a glass microcapillary or a toothpick, spot a small amount of the M&M coloring onto the line drawn on the filter paper. It is best to reapply additional coloring several times to achieve small but easily seen dots.

5. Add enough 1 M NaCl solution to the bottom of a 400-ml beaker or similar jar until there is just enough solution to cover the bottom of the beaker.
6. Carefully place the strip of chromatography paper into the beaker and allow the solvent to migrate by capillary action. It is important that the beaker remain undisturbed during this process.
7. When the solution is about half an inch from the top of the chromatography paper, remove the paper and mark the progress of the solvent using a pencil.
8. Make observations about the individual colors of the M&Ms. Are any of the colors made up of more than one component?

pH DETERMINATION OF EVERYDAY PRODUCTS

Procedure:

PREPARATION OF CABBAGE pH INDICATOR

1. Slice one small red cabbage into small pieces. Place the cut cabbage in a beaker and add distilled water until the cabbage is covered completely with water.
2. Boil gently until the water turns a dark purple color.
3. Set aside to cool.

PREPARATION OF pH STANDARD SOLUTIONS

1. **pH 1 standard**: Pipet about 10 ml of 0.1 M HCl in a 100-ml beaker.
2. **pH 3 standard**: Pipet 1.00 ml of the 0.1 M HCl solution into a 100-ml volumetric flask (can also use a 100-ml graduated cylinder). Fill the flask to exactly 100 ml with distilled water. Pour this solution into a 250-ml beaker. (H^+ concentration = 1.0×10^{-3} M.)
3. **pH 5 standard**: Rinse the volumetric flask several times with water. Pipet 1.00 ml of pH 3 standard solution into the volumetric flask and bring the volume to 100 ml with distilled water. Pour this solution into a 250-ml beaker. (H^+ concentration = 1.0×10^{-5} M.)
4. **pH 7 standard**: Rinse the volumetric flask several times with water. Pipet 1.00 ml of pH 5 standard solution into the volumetric flask and bring the volume to 100 ml with distilled water. Pour this solution into a 250-ml beaker. (H^+ concentration = 1.0×10^{-7} M.)
5. **pH 13 standard**: Pipet about 10 ml of 0.1 M NaOH in a 100-ml beaker. (H^+ concentration = 1.0×10^{-13} M.)
6. **pH 11 standard**: Pipet 1.00 ml of the 0.1 M NaOH solution into a 100-ml volumetric flask. Fill the flask to exactly 100 ml with distilled water. Pour this solution into a 250-ml beaker. (H^+ concentration = 1.0×10^{-11} M.)

7. **pH 9 standard**: Rinse the volumetric flask several times with water. Pipet 1.00 ml of pH 11 standard solution into the volumetric flask and bring the volume to 100 ml with distilled water. Pour this solution into a 250-ml beaker. (H^+ concentration $= 1.0 \times 10^{-9}$ M.)

CREATION OF pH INDICATOR CODE

1. Put about 1 to 2 ml of each of the pH solutions into each of seven test tubes (labeled pH 1, 3, 5, 7, 9, 11, and 13).
2. Add several drops (~15) of red cabbage indicator to each tube.
3. Record the color for each pH standard. The color code you create can now be used to determine the pH of unknown solutions.

pH DETERMINATION OF EVERYDAY PRODUCTS

1. To test the pH of the everyday product solutions, place about 1 to 2 ml of the product in a test tube.
2. Add several drops of the cabbage indicator as above.
3. Record the approximate pH based on the colors obtained using the code you created using the standard pH solutions.
4. Determine the pH of each of the products by placing a drop of the product on the pH paper.
5. Record the pH of the product obtained using the pH paper.
6. Repeat the procedure with each product.

Results:

INDICATOR COLOR

	Color of solution with cabbage indicator
pH 1	
pH 3	
pH 5	
pH 7	
pH 9	
pH 11	
pH 13	

EVERYDAY PRODUCTS

Everyday product	pH determined from indicator	pH determined by pH paper
Soda (clear)		
Vinegar		
Floor cleaner		
Drain cleaner		
Toilet drain cleaner		
Lemon juice		
Clear shampoo		
Ammonia		

Questions:

Did the results obtained using your cabbage indicator agree with that determined with the pH paper?

_____

_____

Which was the better indicator of pH? Why?

_____

_____

TESTING AN HERBICIDE'S EFFECT ON PHOTOSYNTHESIS

Procedure:

INTRODUCTION

Plants use the process of photosynthesis to obtain energy. During photosynthesis, electrons are moved from molecule to molecule. In this laboratory exercise, we will use a chemical dye called DPIP. The DPIP dye turns from a blue color to clear when it gains electrons. The amount of color lost is measured in a spectrophotometer as an increase in the percentage of transmittance.

SETUP

1. Turn on the spectrophotometer and allow it to warm up while you prepare your solutions.
2. Set the wavelength of the spectrophotometer to 605 nm.

3. Label six test tubes on the top rim with the numbers 1 through 6. Wipe the outside of each with a Kimwipe.
4. Prepare test tube 2 by covering the walls and bottom of the tube with foil. Make a foil cap cover for the top. You'll need to remove the foil cover to make your measurements, so make sure you can easily remove and replace the covering. Light should not be permitted inside this tube while it is covered with the foil.
5. Using the table below, set up each tube, **but do not add the unboiled or boiled chloroplasts**.

	1 Blank	2 Unboiled chloroplasts, dark	3 Unboiled chloroplasts, light	4 Boiled chloroplasts, light	5 No chloroplasts (control)	6 Unboiled chloroplasts, herbicide
Buffer	1 ml	1 ml	1 ml	1 ml	1 ml	1 ml
Water	4 ml	3 ml	3 ml	3 ml	3 ml +3 drops	3 ml
DPIP	None	1 ml	1 ml	1 ml	1 ml	1 ml
Unboiled chloroplasts	3 drops	3 drops	3 drops	None	None	3 drops
Boiled chloroplasts	None	None	None	3 drops	None	None
Herbicide	None	None	None	None	None	3 drops

BLANKING THE SPECTROPHOTOMETER

1. Set the spectrophotometer to 0% transmittance with the tube chamber empty by turning.
2. Add the 3 drops of unboiled chloroplasts to tube 1. Cover the top of the tube and invert to mix. (Wipe the tube again with a Kimwipe before inserting into the spectrophotometer.)
3. Insert this tube in the chamber and set the transmittance to 100% using the appropriate knob. Tube 1 is the blank and is used to rezero the spectrophotometer between readings. You must blank/rezero the instrument before each new reading.

MEASUREMENT OF PHOTOSYNTHESIS

1. Stir the *unboiled* chloroplasts with the transfer pipet. Transfer 3 drops of the chloroplast solution to tube 2. Immediately cover the top of the tube, mix, and remove the tube from the foil sleeve.
2. Measure the percentage transmittance of the tube using the blanked spectrophotometer.
3. Record this reading as time 0 for tube 2. Mark the time the tube was read.

4. Replace the tube in the foil sleeve and cover the top with the foil cap.

5. Place the tube in front of the lamp.

6. **You'll need to take new measurements of tube 2 again after it has been exposed to the light from the lamp for 5, 10, and 15 minutes.**

7. Stir the *unboiled* chloroplasts with the transfer pipet. Transfer 3 drops to tube 3. Immediately cover with Parafilm and mix the tube.

8. Measure the percentage transmittance of the tube using the blanked spectrophotometer.

9. Record this reading as time 0 for tube 3. Mark the time the tube was read.

10. Place the tube in front of the lamp.

11. **You'll need to take new measurements of tube 3 again after it has been exposed to the light from the lamp for 5, 10, and 15 minutes.**

12. Stir the *boiled* chloroplast sample, and transfer 3 drops to tube 4. Cover and mix.

13. Measure the percentage transmittance of the tube using the blanked spectrophotometer.

14. Record this reading as time 0 for tube 4. Mark the time the tube was read.

15. Place the tube in front of the lamp.

16. **You'll need to take new measurements of tube 4 again after it has been exposed to the light from the lamp for 5, 10, and 15 minutes.**

17. Cover and mix tube 5 (control, no chloroplasts).

18. Measure the percentage transmittance of the tube using the blanked spectrophotometer.

19. Record this reading as time 0 for tube 5. Mark the time the tube was read.

20. Place the tube in front of the lamp.

21. **You'll need to take new measurements of tube 5 again after it has been exposed to the light from the lamp for 5, 10, and 15 minutes.**

22. Stir the *unboiled* chloroplasts with the transfer pipet. Transfer 3 drops to tube 6. Immediately cover with Parafilm and mix.

23. Measure the percentage transmittance of the tube using the blanked spectrophotometer.

24. Record this reading as time 0 for tube 6. Mark the time the tube was read.

25. Place the tube in front of the lamp.

26. **You'll need to take new measurements of tube 6 again after it has been exposed to the light from the lamp for 5, 10, and 15 minutes.**

Results:

PERCENTAGE TRANSMITTANCE MEASURED OVER TIME

Tube	Time (minutes)			
	0	5	10	15
2 Unboiled, dark				
3 Unboiled, light				
4 Boiled, light				
5 No chloroplasts				
6 Unboiled, herbicide				

Questions:

In which tube did the greatest amount of photosynthesis occur?

_____

_____

Is exposure to light necessary for photosynthesis?

_____

_____

What was the effect of boiling the chloroplasts?

_____

_____

What was the effect of using the herbicide?

_____

_____

TESTING ANTACID EFFECTIVENESS

Procedure:

1. Obtain two small beakers and add 50 ml of distilled water to each.
2. Add 5 drops of the phenol red dye indicator to each beaker and mix.
   This dye will turn from orange-red to yellow in acid solution and
   from orange-red to purple in basic solution.
   What color is the solution?
   What can you conclude about the pH of the solution?
3. Select two different types of antacid (e.g., magnesium- versus calcium-
   based antacid products).
4. Measure 1 ml of a selected antacid and measure its weight.
5. Add the 1 ml of antacid liquid to an individual beaker.
6. Stir each solution.
   What is the color of each solution?
   What does this indicate about the pH of the solution?
7. Slowly add 0.1 M HCl one drop at a time to the beaker, stirring
   constantly.
8. Count and record the total number of drops required to return the
   solution to the original color. (Step 2.)
9. Calculate the number of drops per gram of antacid.
10. Repeat steps 4 to 9 with the second antacid.

Results:

	Antacid 1	Antacid 2
Weight of antacid		
Color of solution after antacid was added to water solution		
Relative pH		
Number of drops needed to neutralize antacid solution		
Number of drops required per gram of antacid		

Glossary

Acetylcholine A chemical found in vertebrate neurons that carries information across the space between two nerve cells. This neurotransmitter binds with two types of receptors, nicotinic and muscarinic.

Alveolus A tiny thin-walled sac located in the lungs rich with capillaries for the exchange of oxygen and carbon dioxide. It is estimated that each human lung has more than 300 million alveoli.

Aminophenol An organic compound with the molecular formula $C_6H_4NH_2OH$. The compound consists of a benzene ring with a hydroxyl (OH) and an amine (NH_2) group attached to it at various positions along the ring. Aminophenols are commonly used as dye intermediates.

Analgesic A medication that reduces or eliminates the sensation of pain.

Anion An ion or group of ions that maintains a negative charge. Anions move toward the positive electrode (anode).

ANS [Autonomic Nervous System] Part of the nervous system of vertebrates that controls involuntary actions of the smooth muscles, heart, and glands. The components of the peripheral and central nervous system concerned with the regulation of involuntary actions. It has two divisions: the sympathetic and parasympathetic nervous systems.

Anticholinergic Inhibiting or blocking the physiological action of acetylcholine at a receptor site. An agent that blocks the parasympathetic nerves.

Antimuscarinic Inhibiting or blocking the action of muscarine or the effects of parasympathetic stimulation at the synaptic junction onto muscles and secretory cells.

Antioxidant A chemical compound or substance that is so easily oxidized that it inhibits the oxidation of other chemical compounds or substances. A substance, such as vitamin E, vitamin C, or β-carotene, thought to protect body cells from the damaging effects of oxidation.

Antipyretic A substance that reduces fever.

BHT [Butylated Hydroxytoluene] 2,6-Bis(1,1-dimethylethyl)-4-methylphenol; 2,6-di-tert-butyl-*p*-cresol; 2,6-di-tert-butyl-4-methylphenol. A chemical compound used as an antioxidant for food, animal feed, petroleum products, synthetic rubbers, plastics, animal and vegetable oils, and soaps. Also found as an antiskinning agent in paints and inks.

Bronchi The main branches of the trachea that lead to the lungs.

Bronchioles Tiny branches that connect the bronchi to the aveoli.

Catalyst A substance that increases the rate of a chemical reaction without being used up in the process. A substance that reduces the activation energy of a reaction, thus increasing the overall rate of the chemical reaction.

Cation An ion or group of ions that maintains a positive charge. Cations move toward the negative electrode (cathode).

CFC [Chlorofluorocarbon] A hydrocarbon that contains chlorine (Cl) and fluorine (F). Once commonly used in aerosol propellants (CFC-11, CFC-12, CFC-113), air conditioning, refrigeration (CFC-12), blowing agents for making foam (CFC-11, CFC-12), cleaning fluids (CFC-113), and solvents for the electronics industry, bedding, and packaging. Chlorofluorocarbons are believed to cause depletion of the atmospheric ozone layer.

CNS [Central Nervous System] The brain and spinal cord of vertebrates.

COX-1 [Cyclo-oxygenase-1] An enzyme that catalyzes the synthesis of prostaglandins from the polyunsaturated fatty acid arachidonic acid. COX-1 is responsible for baseline levels of prostaglandins. Cyclo-oxygenase activity is inhibited by aspirin-like drugs, accounting for their anti-inflammatory effects.

COX-2 [Cyclo-oxygenase-2] An enzyme that catalyzes the synthesis of prostaglandins from the polyunsaturated fatty acid arachidonic acid. COX-2 is responsible for the production of prostaglandins attributable to stimulation (i.e., levels are induced in times of inflammation). Cyclo-oxygenase activity is inhibited by aspirin-like drugs, accounting for their anti-inflammatory effects.

Cytokines Small regulatory proteins that are secreted by immune cells. The proteins act by binding to specific cell receptors, inducing changes in the cell (e.g., stimulating the action of an immune response).

Emollient An agent that softens or soothes the skin. Emollients may be used as lubricants to treat or prevent dry, itchy skin and minor skin irritations.

Emulsifier A surface-active agent that promotes the formation of an emulsion.

Emulsion A suspension of small globules of one liquid in a second liquid with which the first will not mix (e.g., an emulsion of oil and water). Any liquid preparation of a color and consistency resembling milk.

Enzymes Proteins or conjugated proteins produced by living organisms that function as catalysts in specific biochemical reactions.

Histamine A chemical found in animal tissue that is released from mast cells as part of an allergic reaction. It is responsible for the initial symptoms of anaphylaxis: stimulation of gastric secretions, dilation of capillaries, constriction of bronchial smooth muscles, and reduction of blood pressure.

Humectant A chemical compound or substance that promotes the retention of moisture.

Hydrocarbon An organic compound that only contains carbon (C) and hydrogen (H).

Hydrophilic Having a high affinity for water. A substance that tends to dissolve in, mix with, or be soaked by water. Hydrophilic materials tend to be polar in nature.

Hydrophobic Having a lack of affinity for water. A substance that tends not to dissolve in, mix with, or be wetted by water. Hydrophobic materials tend to be nonpolar in nature.

Isomers Compounds that have the same molecular formula but have different structural formulas and thus different properties.

Kaolin A fine clay (China clay; essentially a hydrated aluminum silicate). Common usages include porcelain, ceramics, heat-resistant mortar, clarifying liquids, drying and emollient agents, and as filler or coating for paper and textiles.

Medulla Oblongata The lowermost portion of the vertebrate brain, continuous with the spinal cord, responsible for the control of respiration, circulation, swallowing, and certain other bodily functions. It is the most vital part of the brain in that it controls autonomic functions and relays nerve signals between the brain and the spinal cord.

Nonpolar Molecule A molecule that does not have a positive and a negative end.

NSAIDs [Nonsteroidal Anti-Inflammatory Drugs] Drugs that reduce swelling/inflammation that do not contain any steroid or steroid-like chemicals. Examples include aspirin and ibuprofen. Drugs that reduce swelling/inflammation that do not contain any steroid or steroid-like chemicals.

Opacifying Agent A chemical compound or substance added to a material to make it opaque (i.e., nontransparent).

OTC [Over the Counter] Abbreviation for drugs that can be purchased legally without a medical doctor's prescription.

Oxidation The process of oxidizing. The addition of oxygen to a compound with a loss of electrons; this is always accompanied by reduction.

Parasympathetic Nervous System The part of the peripheral autonomic nervous system originating in the brain stem and the lower part of the spinal cord that, in general, inhibits or opposes the physiological effects of the sympathetic nervous system, as in tending to stimulate digestive secretions, slow the heart, constrict the pupils, and dilate the blood vessels. It is composed of cholinergic (synaptic transmissions are mediated by the release of acetylcholine) ganglion cells located near the target organs.

PEG [Polyethylene Glycol] A water-soluble, waxy solid. The general chemical formula of PEG is $H(OCH_2CH_2)nOH$, where n is greater than or equal to 4. In general, each PEG is followed by a number that corresponds to its average molecular weight. As the molecular weight of PEG increases, viscosity and freezing point increase. PEG is commonly used as a water-soluble lubricant for rubber molds, textile fibers, and metal-forming operations; in food and food packaging; in detergents and as emulsifiers and plasticizers; in water paints, paper coatings, and polishes; and in the ceramics industry. Other common names include macrogol, Carbowax (Union Carbide), Pluracol E (BASF), Poly-G (Olin), and Polyglycol E (Dow).

PG [Prostaglandin] A group of hormone-like substances found in virtually all tissues and organs that are derived from unsaturated fatty acids. These substances mediate a wide range of physiological functions, the most common of which are muscular constriction and the mediation of inflammation. Letters after PG in the abbreviation symbolize different substitutions made to the hydrocarbon backbone, whereas subscript numbers indicate the degree of unsaturation (i.e., the number of double bonds).

Plasticizer A chemical substance added to some plastics to make them more flexible and easier to manipulate.

PNS [Peripheral Nervous System] All of the nerves and neurons that are positioned outside of the brain and spinal cord.

Polar Molecule A molecule that has a positive end and a negative end. This charge can exist as a partial or full charge quantity.

Polymer A molecule with a relatively large molecular weight composed of a series of repeating smaller single (monomer) units.

PPG [Polypropylene Glycol] Abbreviated to the acronym PPG by the International Nomenclature Cosmetic Ingredient. Polypropylene glycol polymers are named as PPG-x, where x is the average number of propylene oxide (C_3H_6O) monomer units (e.g., PPG-10). Esters and ethers of polypropylene glycol polymers are named as PPG derivatives (e.g., PPG-10 stearate, PPG-10 lauryl ether). These compounds are generally used as emollients, emulsifiers, humectants, or surfactants.

Preservative Spoilage deterrent. A chemical compound that is added to protect against decay or decomposition.

Reduction A decrease in positive valence or an increase in negative valence by the gaining of electrons. A reaction in which hydrogen is combined with a compound. A reaction in which oxygen is removed from a compound.

Surfactant A substance that stabilizes a nonpolar substance such as oil in a polar solvent such as water. The surfactant acts by reducing the surface tension of a liquid in which it is dissolved, allowing substances to mix that normally would not mix.

Sympathetic Nervous System The part of the peripheral autonomic nervous system originating in the thoracic and lumbar regions of the spinal cord that, in general, inhibits or opposes the physiological effects of the parasympathetic nervous system, as in tending to reduce digestive secretions, speeding up the heart, and contracting blood vessels. It is composed for the most part of adrenergic ganglion cells (synaptic transmissions are mediated by the release of norepinephrine or epinephrine) located relatively far from the target organs.

Virus A self-replicating, infectious parasite of plants, animals, and bacteria that often causes disease. Typically composed of a nucleic acid (DNA or RNA) protein complex that requires an intact host to replicate.

Index

About the Authors

JOHN TOEDT is Assistant Professor of Physical Sciences at Eastern Connecticut State University where he is also Director of the Biochemistry Program.

DARRELL KOZA is Assistant Professor of Chemistry and serves as the Director of Chemistry for the Center for Antimicrobial Resistance at Eastern Connecticut State University.

KATHLEEN VAN CLEEF-TOEDT is currently Lecturer in the Department of Physiology and Neurobiology at the University of Connecticut.